Developing a Toolkit for Fostering OPEN SCIENCE PRACTICES

PROCEEDINGS OF A WORKSHOP

Thomas Arrison, Jennifer Saunders, and Emi Kameyama, *Rapporteurs*

Committee on Developing a Toolkit for Fostering Open Science Practices: A Workshop

Board on Research Data and Information

Policy and Global Affairs

The National Academies of
SCIENCES · ENGINEERING · MEDICINE

THE NATIONAL ACADEMIES PRESS
Washington, DC
www.nap.edu

THE NATIONAL ACADEMIES PRESS 500 Fifth Street, NW Washington, DC 20001

This workshop activity was supported by the Arcadia Fund under award number 4186, Open Research Funders Group, Open Society Foundations under award number OR2018-45885, Robert Wood Johnson Foundation under award number 78673, and the Wellcome Trust. Any opinions, findings, conclusions, or recommendations expressed in this publication do not necessarily reflect the views of any organization or agency that provided support for the project.

International Standard Book Number-13: 978-0-309-09361-3
International Standard Book Number-10: 0-309-09361-9
Digital Object Identifier: https://doi.org/10.17226/26308

Additional copies of this publication are available from the National Academies Press, 500 Fifth Street, NW, Keck 360, Washington, DC 20001; (800) 624-6242 or (202) 334-3313; http://www.nap.edu.

Copyright 2021 by the National Academy of Sciences.

This work is licensed under the Creative Commons Attribution 4.0 International License (CC BY 4.0). To view a copy of the license, visit https://creativecommons.org/licenses/by/4.0/.

Printed in the United States of America

Suggested citation: National Academies of Sciences, Engineering, and Medicine. 2021. *Developing a Toolkit for Fostering Open Science Practices: Proceedings of a Workshop*. Washington, DC: The National Academies Press. https://doi.org/10.17226/26308.

The National Academies of
SCIENCES · ENGINEERING · MEDICINE

The **National Academy of Sciences** was established in 1863 by an Act of Congress, signed by President Lincoln, as a private, nongovernmental institution to advise the nation on issues related to science and technology. Members are elected by their peers for outstanding contributions to research. Dr. Marcia McNutt is president.

The **National Academy of Engineering** was established in 1964 under the charter of the National Academy of Sciences to bring the practices of engineering to advising the nation. Members are elected by their peers for extraordinary contributions to engineering. Dr. John L. Anderson is president.

The **National Academy of Medicine** (formerly the Institute of Medicine) was established in 1970 under the charter of the National Academy of Sciences to advise the nation on medical and health issues. Members are elected by their peers for distinguished contributions to medicine and health. Dr. Victor J. Dzau is president.

The three Academies work together as the **National Academies of Sciences, Engineering, and Medicine** to provide independent, objective analysis and advice to the nation and conduct other activities to solve complex problems and inform public policy decisions. The National Academies also encourage education and research, recognize outstanding contributions to knowledge, and increase public understanding in matters of science, engineering, and medicine.

Learn more about the National Academies of Sciences, Engineering, and Medicine at **www.nationalacademies.org**.

The National Academies of
SCIENCES • ENGINEERING • MEDICINE

Consensus Study Reports published by the National Academies of Sciences, Engineering, and Medicine document the evidence-based consensus on the study's statement of task by an authoring committee of experts. Reports typically include findings, conclusions, and recommendations based on information gathered by the committee and the committee's deliberations. Each report has been subjected to a rigorous and independent peer-review process and it represents the position of the National Academies on the statement of task.

Proceedings published by the National Academies of Sciences, Engineering, and Medicine chronicle the presentations and discussions at a workshop, symposium, or other event convened by the National Academies. The statements and opinions contained in proceedings are those of the participants and are not endorsed by other participants, the planning committee, or the National Academies.

For information about other products and activities of the National Academies, please visit www.nationalacademies.org/about/whatwedo.

PLANNING COMMITTEE ON DEVELOPING A TOOLKIT FOR FOSTERING OPEN SCIENCE PRACTICES

Loretta Parham (*Chair*), Chief Executive Officer and Library Director, Robert W. Woodruff Library, Atlanta University Center
Stuart Buck, Former Vice President of Research, Arnold Ventures
Sarah Nusser, Professor, Department of Statistics, Iowa State University
Roger Wakimoto, Vice Chancellor for Research, University of California, Los Angeles

Project Staff

Thomas Arrison, Director, Board on Research Data and Information
George Strawn, Scholar, Board on Research Data and Information
Ester Sztein, Deputy Director, Board on Research Data and Information
Emi Kameyama, Program Officer, Board on Research Data and Information

Consultants

Jennifer Saunders, Consultant
Greg Tananbaum, Director, Open Research Funders Group

ROUNDTABLE ON ALIGNING INCENTIVES FOR OPEN SCIENCE

Thomas Kalil (*Co-Chair*), Chief Innovation Officer, Schmidt Futures
Keith Yamamoto (NAS/NAM) (*Co-Chair*), Vice Chancellor for Science Policy and Strategy, University of California, San Francisco
Elizabeth Albro, Commissioner, National Center for Education Research, U.S. Department of Education*
Danny Anderson, President, Trinity University
Roslyn Clark Artis, President, Benedict College
Chris Bourg, Director of Libraries, Massachusetts Institute of Technology
Courtney Brown, Vice President for Strategic Impact, Lumina Foundation*
Anne-Marie Coriat, Head, UK and Europe Research Landscape, Wellcome Trust*
Michael Crow, President, Arizona State University
Mark Cullen (NAM), Director, Center for Population Health Sciences; Senior Associate Dean for Research, School of Medicine, Stanford University
Ronald Daniels, President, Johns Hopkins University
Tashni-Ann Dubroy, Executive Vice President and Chief Operations Officer, Howard University
Susan Fitzpatrick, President, James S. McDonnell Foundation*
Maryrose Franko, Executive Director, Health Research Alliance*
Nicholas Gibson, Senior Program Officer, Human Sciences, John Templeton Foundation*
Daniel Goroff, Vice President and Program Director, Alfred P. Sloan Foundation*
Heide Hackmann, Chief Executive Officer, International Science Council*
Randolph Hall, Vice President for Research and Professor, University of Southern California
Robert Hanisch, Director, Office of Data and Informatics, National Institute of Standards and Technology*

---

*Denotes ex officio member.

Patricia Hswe, Program Officer for Scholarly Communications, Andrew W. Mellon Foundation*
Adam Jones, Program Officer, Science Program, Gordon and Betty Moore Foundation*
Renu Khator, President, University of Houston
Richard McCullough, Vice Provost for Research, Harvard University
Peter McPherson, President, Association of Public and Land-grant Universities*
Ross Mounce, Director of Open Access Programs, Arcadia Fund*
Sarah Nusser, Professor, Department of Statistics, Iowa State University
Loretta Parham, Chief Executive Officer and Director, Robert W. Woodruff Library, Atlanta University Center
Heather Pierce, Senior Director, Science Policy and Regulatory Counsel, Association of American Medical Colleges*
Dawid Potgieter, Senior Program Officer, Templeton World Charity Foundation*
Brian Quinn, Assistant Vice President, Research-Evaluation-Learning, Robert Wood Johnson Foundation*
Robert Robbins, President, University of Arizona
Jerry Sheehan, Deputy Director, National Library of Medicine, National Institutes of Health*
Bodo Stern, Chief of Strategic Initiatives, Howard Hughes Medical Institute*
Barbara Snyder, President, Association of American Universities*
Geeta Swamy, Associate Vice President for Research, Duke University; Vice Dean for Scientific Integrity, Duke University School of Medicine
Shirley Tilghman (NAS/NAM), President Emerita, Princeton University
Alan Tomkins, Deputy Director, Social, Behavioral and Economic Sciences, National Science Foundation*
Roger Wakimoto, Vice Chancellor for Research, University of California, Los Angeles
Thomas Wang, Chair, Open Science Committee, American Heart Association; Director, Division of Cardiovascular Medicine, Vanderbilt University Medical Center*
Jennifer Weisman, Director and Chief of Staff, Global Health Division, Bill & Melinda Gates Foundation*
Richard Wilder, General Counsel and Director of Business Development, Coalition for Epidemic Preparedness Innovations*

Richard Wilder, General Counsel and Director of Business Development, Coalition for Epidemic Preparedness Innovations*
Duncan Wingham, Executive Chair, Natural Environment Research Council, United Kingdom Research and Innovation*

Open Research Funders Group Secretariat

Heather Joseph, Executive Director, SPARC
Greg Tananbaum, Director, Open Research Funders Group

Staff

Thomas Arrison, Director, Board on Research Data and Information
George Strawn, Scholar, Board on Research Data and Information
Ester Sztein, Deputy Director, Board on Research Data and Information
Emi Kameyama, Program Officer, Board on Research Data and Information
Olivia Torbert, Senior Program Assistant, Board on Research Data and Information

BOARD ON RESEARCH AND DATA INFORMATION

Sarah Nusser (*Chair*), Professor, Department of Statistics, Iowa State University
Amy Brand, Director, MIT Press
Stuart Feldman, Chief Scientist, Schmidt Futures
Ian Foster, Arthur Holly Compton Distinguished Service Professor of Computer Science, The University of Chicago; Director, Data Science and Learning Division, Argonne National Laboratory
Ramanathan Guha, Google Fellow, Google
Sallie Ann Keller (NAE), Division Director, Social and Decision Analytics and Distinguished Professor in Biocomplexity, University of Virginia
Mary Lee Kennedy, Executive Director, Association of Research Libraries
Barend Mons, Chair, BioSemantics, Leiden University Medical Center
Michael Stebbins, President, Science Advisors, LLC
Bonnie Carroll, Retired Founder and Strategic Consultant, Information International Associates, Inc.; CODATA Secretary General*
John Hildebrand (NAS), Regents Professor of Neuroscience, University of Arizona; NAS International Secretary*

Staff

Thomas Arrison, Director, Board on Research Data and Information
George Strawn, Scholar, Board on Research Data and Information
Ester Sztein, Deputy Director, Board on Research Data and Information
Emi Kameyama, Program Officer, Board on Research Data and Information
Olivia Torbert, Senior Program Assistant, Board on Research Data and Information

---

*Denotes ex officio member.

Preface and Acknowledgments

The National Academies of Sciences, Engineering, and Medicine's Roundtable on Aligning Incentives for Open Science convenes critical stakeholders to discuss the effectiveness of current incentives for adopting open science practices, current barriers of all types, and ways to move forward to optimally align reward structures and institutional values. To increase the contribution of open science to producing better science, the roundtable aims to improve coordination among stakeholders and increase awareness of current and future efforts in the broader scientific community. At its first meeting in February 2019, the roundtable developed an initial work plan and set of priorities. On September 20, 2019, the roundtable organized a public symposium, and the resulting publication, *Advancing Open Science Practices: Stakeholder Perspectives on Incentives and Disincentives: Proceedings of a Workshop—in Brief*, was released in 2020. The third meeting of the roundtable was held on February 27, 2020. Six working groups, created following the inaugural meeting, are driving the work of the roundtable to define specific issues or problems related to open science and discuss possible actions.

An important focus of roundtable discussions and working group activities to date has been the need for information and other resources that could be used by researchers, research institutions, research funders, professional societies, and other stakeholders interested in fostering open science practices. For example, universities that are considering adopting new approaches to encourage open science practices might benefit from example cover letters or essays introducing open science-related issues that could be adapted for use in their communities. Research institutions and research funders could benefit from examples of language that signals their interest in open science activities and could be utilized in grant applications or job postings. Good practices primers that point to existing open science-related policies and approaches could be useful to a range of stakeholders. A compilation of the various research outputs that stakeholders might consider in developing approaches that foster open science could be useful in cross-disciplinary discussions, since different disciplines generate different types of outputs.

On November 5, 2020, the Roundtable on Aligning Incentives for Open Science organized a virtual public workshop in conjunction with its fourth meeting on November 6. The workshop explored the information and resource needs of researchers, research institutions, research funders, professional societies, and other stakeholders interested in fostering open science practices. Workshop participants discussed approaches to meeting those needs, such as development of a toolkit that could be used by various groups of stakeholders. The workshop included presentations of commissioned papers that describe and provide examples of draft elements of a toolkit that have been developed by members of roundtable working groups (see Appendix C). The workshop and these proceedings were made possible by financial support from the Arcadia Fund, the Open Research Funders Group, the Open Society Foundations, the Robert Wood Johnson Foundation, and the Wellcome Trust. Over its first three years, the roundtable has also received support from Arnold Ventures, Schmidt Futures, the Leona M. and Harry B. Helmsley Charitable Trust, and the National Library of Medicine.

This Proceedings of a Workshop was prepared by the workshop rapporteurs as a factual summary of what was presented and discussed at the workshop. The planning committee's role was limited to planning and convening the workshop. The statements made are those of the rapporteurs and do not necessarily represent positions of the workshop participants as a whole, the planning committee, or the National Academies of Sciences, Engineering, and Medicine. We wish to extend sincere thanks to all the members of the planning committee for their contributions in scoping, developing, and carrying out this project.

This proceedings has been reviewed in draft form by individuals chosen for their diverse perspectives and technical expertise, in accordance with procedures approved by the National Academies of Sciences, Engineering, and Medicine. The purpose of this independent review is to provide candid and critical comments to assist the institution in making its published report as sound as possible and ensure the document meets institutional standards for quality and objectivity. The review comments and draft manuscript remain confidential to protect the integrity of the process. We wish to thank the following individuals for their review of this proceedings: Daniel Himmelstein, University of Pennsylvania; Veronique Kiermer, Public Library of Science; Alexa McCray, Harvard Medical School; and Keith Webster, Carnegie Mellon University. Although the reviewers listed above have provided many constructive comments and suggestions, they were not asked to endorse the content of the proceedings, nor did they see the final draft before its release. The review of this proceedings was overseen by Mary Lee Kennedy, Association of Research Libraries. Appointed by the National Academies, she was responsible for making certain that an independent examination of this report was carried out in accordance with institutional procedures and that all review comments were carefully considered. Responsibility for the final content of this proceedings rests entirely with the rapporteurs and the National Academies.

Thomas Arrison, Director
Board on Research Data and Information
National Academies of Sciences, Engineering, and Medicine

Contents

1	INTRODUCTION	1
2	ADOPTING AND UTILIZING A TOOLKIT FOR OPEN SCIENCE: STAKEHOLDER PERSPECTIVES	5
3	ROUNDTABLE PRIORITIES FOR ADVANCING OPEN SCIENCE	15
	REFERENCES	21
	APPENDIXES	
A	WORKSHOP AGENDA	23
B	BIOGRAPHIES OF SPEAKERS AND MODERATORS	25
C	TOOLKIT ELEMENTS	31

1

Introduction

The movement toward open science has facilitated the dissemination of new information and insights and has spurred innovation. The National Academies of Sciences, Engineering, and Medicine defines that open science "aims to ensure the free availability and usability of scholarly publications, the data that result from scholarly research, and the methodologies, including code or algorithms that were used to generate those data" (NASEM, 2018). While open science has the potential to close knowledge gaps and level the playing field for researchers in the United States and globally, there are barriers to bringing it to scale for broader use, particularly for those who carry out the research. Universities, funding agencies, societies, philanthropies, and industry all face varying challenges in implementing open science practices supportive of a rigorous, transparent, and effective research culture.

The National Academies Roundtable on Aligning Incentives for Open Science, established in 2019, has taken on an important role in addressing issues with open science. The roundtable convenes critical stakeholders to discuss the effectiveness of current incentives for adopting open science practices, current barriers of all types, and ways to move forward in order to align reward structures and institutional values. The roundtable convenes two times per year and creates a venue for the exchange of ideas and joint strategic planning among key stakeholders including universities, funding agencies, societies, philanthropies, and industry whose organizations have ambitious missions. The roundtable aims to improve coordination among stakeholders and increase awareness of current and future efforts in the broader open science community, in an attempt to properly incentivize a more rigorous, transparent, and effective research culture.

An important focus of the roundtable's work to date has been defining resources that can help key stakeholders discuss, develop, and deploy open science incentivization plans that are both consistent with common norms and appropriate for their specific communities. Given the points of leverage that these stakeholders manage (e.g., hiring, review, tenure and promotion, and funding), how can they be activated to create better alignment across research values, practices, and incentives? Put succinctly, incentives are the tools we use to ensure that research practices are consistent with the organizational values we espouse (see Figure 1-1).

WORKSHOP INTRODUCTION, CONTEXT, AND ORGANIZATION

To further elucidate some of these issues, a virtual public workshop on fostering open science practices was convened on November 5, 2020, in conjunction with the fall 2020 meeting of the roundtable. The broad goal of the workshop was to identify paths to growing the nascent coalition of

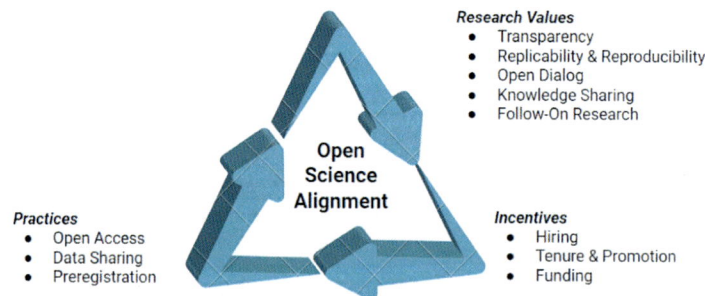

FIGURE 1-1 Open science alignment.
SOURCE: Author generated.
NOTE: The bullet points are examples to highlight a subset of the elements in each of the three categories that the Roundtable on Aligning Incentives for Open Science focused on.

stakeholders committed to reenvisioning credit/reward systems (e.g., academic hiring, tenure and promotion, and grants) to fully incentivize open science practices. The workshop explored the information and resource needs of researchers, research institutions, government agencies, philanthropies, professional societies, and other stakeholders interested in further supporting and implementing open science practices (described further below).

The workshop also included a broad group of thought leaders and researchers to share perspectives on adopting and utilizing information resources compiled in the form of a toolkit. As background for the discussion, several toolkit elements developed by expert authors were circulated in advance of the workshop to stimulate discussion among the community about how such a toolkit might be used, what additional materials are needed, and how such a toolkit should be disseminated for broad adoption. The toolkit is primarily intended to assist university leadership, academic department chairs, research funders, learned societies, and government agencies; revised toolkit elements are included in Appendix C.

After framing remarks (summarized below), the agenda was organized in several sessions; this workshop proceedings follows the organization of the workshop. Chapter 2 summarizes presentations from various perspectives (university, disciplinary, and research funder) on adopting and utilizing the toolkit for open science. The session that followed included breakout discussions where participants were asked to identify priorities for the roundtable to bring incentive structures into alignment with open science practices. These discussions are summarized in Chapter 3 along with closing remarks. The appendixes include the agenda and biographical sketches of committee members and presenters (Appendixes A and B, respectively). The draft toolkit is in Appendix C.

INTRODUCTORY AND FRAMING REMARKS

Keith Yamamoto, co-chair of the Roundtable on Aligning Incentives for Open Science and vice chancellor for science policy and strategy of the University of California, San Francisco, opened the workshop with an introduction and framing remarks. He summarized the goal and purpose of the workshop, which is to bring together a wide range of stakeholders to discuss, develop, and deploy open science tools appropriate for specific stakeholder communities (i.e., university, disciplinary, and researcher funder) and to develop effective points of leverage for each of the stakeholder groups.

Yamamoto stated that workshop participants will discuss how best to identify and nurture an environment to support, encourage, recognize, and reward open science practices.

As described above, Yamamoto noted that the workshop will inform the development of a toolkit to promote open science practices. As part of this discussion, he noted that consideration of incentives is an important focus, including, for example, adopting and adapting signaling language to highlight an organization's interest in open science activities and specific leverage points, such as grant applications or job postings.

Loretta Parham, chair of the workshop planning committee and chief executive officer and library director of the Robert W. Woodruff Library of the Atlanta University Center, offered introductory and welcome remarks to participants, adding that the goal of the workshop is to advance the movement to make open science the norm.

Thomas Kalil, co-chair of the Roundtable on Aligning Incentives for Open Science and chief innovation officer of Schmidt Futures, discussed the goal of the breakout sessions, which is to build a coalition of stakeholders engaged and invested in creating powerful incentives for open science. Breakout session participants will identify actions they can individually take to create these incentives, and they will identify gaps where more work is needed.

Kalil highlighted the importance of professional societies in this issue, describing the example of the Council of Graduate Departments of Psychology, which has been moving the field forward by creating resources for department chairs to encourage open science, such as research training methods and draft tenure and promotion language. Similarly, the provost of the Massachusetts Institute of Technology issued a call for each department laboratory and center to create local plans to encourage and support open sharing.

Kalil added that while these examples are encouraging, given the many types of data and publications that are produced across different disciplines, efforts to support open science across these disciplines will require an all-hands approach. Breakout sessions provide an opportunity for participants to reflect and share actions they can take as part of a larger effort to move this area forward.

2

Adopting and Utilizing a Toolkit for Open Science: Stakeholder Perspectives

A key goal of the workshop was to gather input on advancing open science from various key stakeholders, including university, disciplinary, and research funder perspectives. Presentations from individuals representing these various perspectives are summarized below.

UNIVERSITY PERSPECTIVES

During this session, speakers provided a university perspective related to fostering open science practices.

Michael Crow, president of Arizona State University (ASU), opened the session, providing his experience as president of the university and as deputy provost and chief research officer at Columbia University. During his time in these positions, he noted that communicating about science, particularly with the public, has been a significant challenge, slowing progress in advancing open science.

One of the challenges in the movement toward open science has been the bureaucratic nature of academia, noted Crow. "We will never have open science at the level that we need … until we have open organizational structures and open organizational design," he stated, adding that open science is the tool needed to support exploration and innovation.

However, there are examples of successful moves toward open science in academia. He cited ASU's new School of Sustainability, which connects a wide range of topics and supports open science. ASU has 600 faculty members involved in initiatives connecting supply chain dynamics to computational assessment, biology, and complexity theory, all working together to address critical challenges and share data and resources.

Crow added that university administration needs to understand that openness is more than just the "opening up" of scientific trajectory and that it includes the broadening of scientific perspectives, theoretical outcomes, and new theories and practices. It is also the broadening of the notion of the understanding of science by the entire population. What is needed, he said, is the bold redesign of universities and scientific enterprises; these need to open to break down barriers. This shift, which is particularly critical at the leadership level, may also require broadening of funding sources to more nontraditional funders.

Philip Bourne, founding dean of the University of Virginia School of Data Science, described the need for more flexibility among organizational types to achieve open science, noting three entry points (bottom up, middle, and top down) that offer opportunities to change and align incentives for these efforts.

Describing a bottom-up approach, Bourne discussed his work with a biological database, the Protein Data Bank, run by scientists who created a culture and strong incentives to move toward open data sharing. In an example of a middle-ground approach, Bourne discussed his work with the Public Library of Science, or PLOS, a nonprofit open access science, technology, and medicine publisher with a portfolio of open access journals and other content. The organization's incentives for openness included an interest in working with prestigious journals as well as an interest in and desire to be part of contributing to the movement toward open science. Bourne's work with the National Institutes of Health (NIH) provided an example of a top-down model such as the NIH FAIR (Findability, Accessibility, Interoperability, and Reuse of digital assets) data-sharing policy and the policy on preprints. While there are many opportunities to make an impact in open science, he said, change comes slowly.

Bourne noted that there are points of entry for each of the bottom-up, middle, and top-down organizational structures to support open science. For example, he noted that in hiring faculty, universities should work to develop the culture from the ground up, examining how promotion and tenure policies and the faculty handbook can support this culture of openness. In terms of middle-ground incentives, these can come about through partnerships that support open science. Partnerships with libraries are critical to this movement.

Regarding top-down initiatives, Bourne noted that the president, provost, and the chief operating officer of the University of Virginia were involved in the development of a new strategic plan for the institution that supported a movement toward openness. This resulted in an improvement in the visibility of research that comes from open science and data sharing. He added that the main incentive in moving the cultural needle on this issue is partnerships and working with other teams across schools. Stakeholders at all levels within the community need to be engaged in developing a culture of openness—from students through university presidents.

Tatiana Bryant, research librarian for digital humanities, history, and African American studies at the University of California, Irvine, described a current research project in collaboration with Camille Thomas, scholarly communications librarian at Florida State University, to examine Black, Indigenous, and people of color (BIPOC) faculties' experience with open access, including what, if any, institutional and disciplinary support these faculties have received in this area.

Bryant said she and Thomas had planned to attend conferences to convene focus groups across a range of disciplines but had to pivot to online focus groups due to the COVID-19 pandemic. This resulted in eight rounds of focus groups of about 40 participants. Several factors, including difficulty recruiting faculty during the summer months and the disproportionate impact the pandemic has had on BIPOC, made recruiting difficult. Given these challenges, a majority of faculty who have participated in the focus groups represent humanities and social science disciplines, rather than science, technology, engineering, and mathematics fields.

While the study is ongoing, some preliminary data have been collected, stated Thomas. These indicate the need to address barriers to open science at their source. Participants also noted that institutions or departments with a social justice mission appeared to be more encouraging of open science practices. Being considered contingent labor along with the risk of not receiving promotion and tenure were identified as barriers. For example, several participants noted that they were postdocs prior to their current faculty positions; open science initiatives were not available to them as postdocs. Additionally, Thomas noted that initial results indicated that BIPOC faculty tended to be excluded from collaborative networks that received grant funding, particularly federal grant funding, where policies were supportive of open science practices.

Thomas added that universities could further support BIPOC faculty in open science efforts by publishing policies, mandates, and expectations in institutions as well as offering startup and research funds for article processing charges at the individual and faculty level. However, she noted that participants described concerns about predatory publishing, including the need to consider journal quality to address these concerns. Finally, Thomas noted that offering seed money for new projects and open access should be weighted as a part of the application process.

DISCIPLINARY PERSPECTIVES

During this session, speakers provided a disciplinary perspective related to adopting and utilizing a toolkit to promote open science.

Lauren Collister, chair of the Committee on Scholarly Communication in Linguistics of the Linguistic Society of America (LSA), began by providing an overview of open scholarship in linguistics, including current efforts by the LSA related to open science. The field is strongly aligned with open science; sharing coded data, problem sets, and publications is a norm in the field, although it is not always necessarily articulated as open science or open scholarship. With a strong foundation in openness, linguists share knowledge, training, and tools, including, for example, through the LSA's hub as well as at its annual meetings and training seminars. These tools and resources can be used by scholars to advocate for the open practices that underlie their work. The LSA also provides support on how to develop, manage, and openly share data.

The LSA has developed a statement on the evaluation of language documentation for hiring, promotion, and tenure.[1] Department chairs have used this statement as a lever to change practices and incentives for tenure and promotion in their universities. Another statement currently in draft form, developed by the LSA's Committee on Scholarly Communication in Linguistics, addresses the scholarly merit and evaluation of open scholarship.[2]

Collister described how the LSA has worked to move toward open journal publishing, while considering the financial impact this has had on the society. The society has been using a hybrid/delayed open access model for its flagship journal, which has other fully subsidized open access publications. While open access publishing is important, it does not have to be

---

[1] See https://www.linguisticsociety.org/resource/statement-evaluation-language-documentation-hiring-tenure-and-promotion.
[2] See https://www.linguisticsociety.org/content/statement-scholarly-merit-and-evaluation-open-scholarship-linguistics.

the sole focus of a society's work in moving toward open science. Societies have the power to drive change for a discipline; they can use that power to elevate open scholarship, stated Collister (see Figure 2-1). The Roundtable on Aligning Incentives for Open Science's toolkit along with other published guidance on signaling language can support a shift toward openness. Embracing openness and the need for culture change is essential to moving forward, she added.

Sanjay Srivastava, professor and undergraduate education chair of the University of Oregon Department of Psychology, began by noting that the movement toward openness in psychology was initially motivated by concerns about the inability to replicate research; lack of transparency in research methods made it difficult to fully evaluate the evidence. However, the field has long been engaged in open scholarship to address these issues. The University of Oregon, for example, has been incorporating open science into its teaching, both in graduate seminars and in undergraduate classes.

Several years ago, the psychology department at the University of Oregon began incorporating open science principles into its hiring practices, Srivastava said. The department, in an effort to signal the importance of openness to its candidates and faculty, incorporated the following language, adapted from the Ludwig Maximilian University of Munich and the 2016

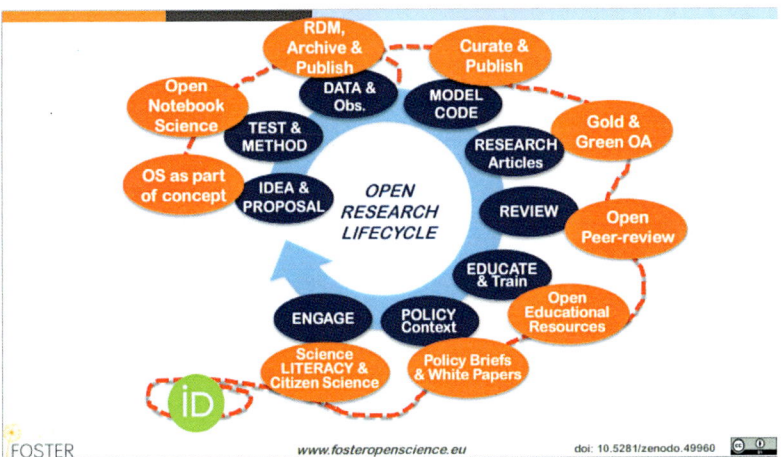

FIGURE 2-1 Open research life cycle.
SOURCE: Lauren Collister, presentation, National Academies of Sciences, Engineering, and Medicine, November 5, 2020, based on www.fosteropenscience.eu.

conference of the Society for the Improvement of Psychological Science, into its job advertisements: "Our department embraces the values of open and reproducible science, and candidates are strongly encouraged to address in their statements how they have pursued and/or plan to pursue these goals in their work." This has encouraged applicants to discuss their interest and experience with open science in their research statements or cover letters. Srivastava added that the department has also had a strong emphasis on inclusion. As the department works to diversify and expand, the goal is to hire broadly to represent the field.

More recently, the department has been working to develop guidance for incorporating openness into tenure and promotion activities. The department held workshops for tenure-track, non-tenure-track, and early-career faculty to identify priorities, valuing, in particular, the early-career perspective. Participants discussed the importance of expanding research outputs beyond papers and publications to address open science. Developing research philosophy statements to incorporate open science was also discussed. Other initiatives include encouraging candidates to nominate a set of representative papers for review. To support this process, external referees would be provided guidance about how to evaluate these papers and consider the candidate's work through this lens.

RESEARCH FUNDER PERSPECTIVES

Speakers during this session provided a research funder's perspective related to increasing open science activities.

Ekemini Riley, managing director of Aligning Science Across Parkinson's (ASAP), described the organization's work to accelerate the pace of discovery for Parkinson's disease specifically and neurodegeneration broadly. ASAP's collaborative research network, one of the organization's key funding initiatives, brings together multiple disciplines, institutions, career stages, and geographies, while aiming to foster collaboration across teams early in the research process. Open science and research transparency are key requirements in receiving funding from ASAP, Riley noted. Funded researchers must agree to the following terms:

- Share new results on the ASAP Hub (virtual platform)
- Post experimental protocols to a protocol-sharing service, such as protocols.io

- Deposit tools (and code) through accessible repositories once the tool has been characterized and validated through the funded study
- Submit manuscripts to a preprint server, such as bioRxiv, before or concurrent to the first submission to a journal
- Publish finished work in a journal that offers immediate open access with a Creative Commons CC BY license

The research network now spans 11 countries. Those on the research team, from principal investigators to graduate students, can view funded proposals that are posted on the research hub, serving as a point for research collaboration.

A movement toward open science policies and practices requires an intentional culture change, including team-centric approaches and data sharing early and often, stated Riley. To achieve results in this area, there is a need for a constellation of open science mechanisms and opportunities for virtual collaboration.

Julia Stewart Lowndes, senior fellow at the National Center for Ecological Analysis and Synthesis of the University of California, Santa Barbara, and founder of Openscapes, began by noting that data analysis is a critical area to focus efforts to expand open science. Current approaches for data analysis are closed and individualized; however, open data science changes this perspective by offering collaborative and empowering opportunities for researchers. Open science is also a powerful tool to embrace diversity and inclusivity.

Lowndes described the Ocean Health Index,[3] a program designed to measure ocean health from a global to local scale at the University of California, Santa Barbara. A 2017 publication summarized the program's efforts to move toward open data science approaches (Lowndes et al., 2017); new skillsets and mindsets were essential to this effort. Open science was identified as a means to meet analytical needs to develop the program. The Ocean Health Index, now used globally, for example, by the United Nations, has demonstrated the ability to reimagine public engagement and science communication through its open access to code, data, and publishing.

Lowndes discussed efforts by Openscapes to help researchers practice open science, by approaching open science as (1) a spectrum, with many

---

[3] See http://ohi-science.org.

entryways to meet researchers where they are; (2) a behavior change, requiring new skillsets and mindsets; and (3) a movement, where empowering leaders and champions is valued along with individual skills.[4] She then described the Openscapes Champions mentorship program offered to research teams. It is a remote-by-design and cohort-based program, enabling community building across research teams and universities (Lowndes, 2019). This program helps to address an unmet need to support research teams through training and mentorship, she said; support for open leaders who are trainers and mentors is also needed.

Lowndes cited a framework developed by the Center for Open Science on how to support a cultural shift toward open science (see Figure 2-2). As a research community, Lowndes said we have made progress to make open science possible, easy, rewarding, and required, in some cases, as part of our fundamental processes. Now, she added, we need to support researchers to make it normative, which requires investing in human infrastructure. This can include sponsoring research teams to be trained and mentored. Researchers can attend hack weeks and meetups and can participate in workshops and as Openscapes champions.

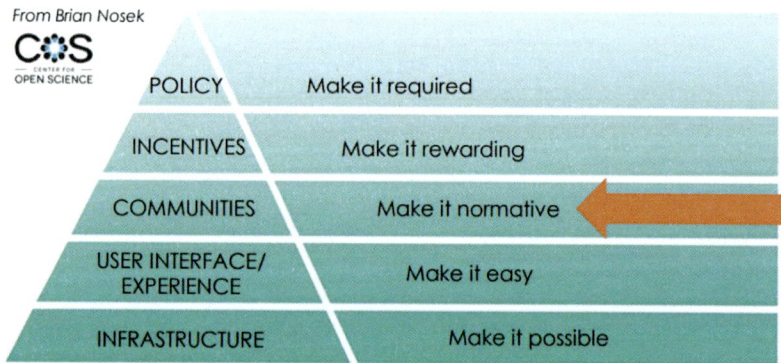

FIGURE 2-2 Changing a research culture.
SOURCE: Julia Stewart Lowndes, presentation, National Academies of Sciences, Engineering, and Medicine, November 5, 2020, based on Brian Nosek, Center for Open Science.

---

[4] See http://openscapes.org.

In addition, Lowndes said, there is a need for salaried positions for open leaders to train and mentor research teams. Trainers, mentors, research software engineers, and community managers who participate in this work are often early-career academics supported by "soft money." Increasing the number of salaried positions in this area would allow for increased continuity in these roles and would reduce burnout. Open leaders engaged in this work are doing this without support and many leave academia as a result, ultimately slowing the progress toward open science. Lowndes noted that open leaders do jobs that are not currently valued in academia, and yet they are critical to normalizing open science and making progress in this area.

3

Roundtable Priorities for Advancing Open Science

Workshop participants broke into several small group sessions with the goal of suggesting possible strategies for the Roundtable on Aligning Incentives for Open Science to catalyze changes to research incentivization systems at scale. Greg Tananbaum of the Open Research Funders Group moderated a session in which a representative of each of the breakout groups summarized the group's discussion. Five priorities emerged from those summaries for the roundtable to consider that could help bring incentive structures into alignment with open science practices:

1. Ensure researchers have the guidance, training, and resources to fully participate in open science practices.
2. Encourage deans, presidents, and provosts to signal that open science is a priority.
3. Make the reporting of open activities and the accrual of credit both easy and normative.
4. Build upon policies and processes already in place to support open science.
5. Identify and address the financial and human costs of open science.

(1) Ensure researchers have the guidance, training, and resources to fully participate in open science practices. Susan Fitzpatrick of the

James S. McDonnell Foundation said that transitioning to a new system of research dissemination carries a degree of uncertainty and risk for all parties. This can be mitigated through proactive measures such as aligning the effort with clear incentives for researchers, leveraging existing institutional infrastructure to socialize and support new norms, and coordinating with funders and professional societies to develop discipline-specific guidance.

The roundtable should work to systematically ensure scientists, especially those working at universities, are aware of the available resources at their disposal, on campus and beyond, said Derrick Anderson of Arizona State University. Researchers can learn from exemplars in their disciplines who are already successfully practicing and championing in open science activities. Michael Dougherty of the University of Maryland discussed the importance of devising new training and professional development programs for students, faculty, and staff on open science tools, methodologies, and approaches as gaps in these available resources and case studies are identified. Fitzpatrick also noted that funding will be necessary support to foster open tools and practices. Giving visibility to institutions and departments providing training in open scholarship and demonstrating the impact could be beneficial as well, said Jennifer Gibson of eLife. Russell Poldrack of Stanford University added that there is a need to develop tools to mentor people at the beginning of their careers to adopt open science practices and give professional recognition for sharing data and preprints.

Participants also discussed assessing and reevaluating the role of open science as part of the faculty workload, including creating incentives and rewards. Robert Hanisch of the National Institute of Standards and Technology noted that faculty are expected to drive change in open science and yet are not recognized for their efforts through the current promotion and tenure process. To address this, Dougherty suggested consistent language in faculty job ads and annual review criteria. This would ensure that researchers are encouraged to pursue open science activities and practices as a core element of their workload. Gibson agreed that there is a need to recognize open science and collaboration or team science in tenure and promotion processes, which could perhaps be accomplished in part by highlighting open data citations on CVs. Maryrose Franko of the Health Research Alliance said providing guidance and examples of wording for applications could help job candidates convey the impact of open science in their work. This language could be highlighted in promotion and tenure packages and grant applications.

(2) Encourage deans, presidents, and provosts to signal that open science is a priority. Breakout session participants emphasized the critical role that senior leadership can play in signaling the importance of open science at their institutions. A concerted effort to bring campus and institutional leaders on board with open science is key if open science is going to be viewed as valuable and rewarded, said Dougherty. One way to promote open science could be to host workshops with academic consortia, such as the Big Ten Academic Alliance.

Boyana Konforti of the Howard Hughes Medical Institute reiterated the importance and role of university leaders embracing open science, noting that broader communication with faculty and administrators at the college level is needed to link the movement with policies. Katie Steen of the Association of American Universities also discussed how university leaders can use high-level signaling statements to communicate the value of open research broadly across campus. Chris Bourg of the Massachusetts Institute of Technology added that leadership can build upon those signals by supporting the infrastructure necessary for open science and scholarship.

University leaders can also play a role responding to structural racism, including connecting this movement to open science policies, said Konforti. Historically Black Colleges and Universities can play a strong role. To this point, Bourg added that leaders can promote a broader understanding of the role open science plays in enabling research excellence and inclusivity.

(3) Make the reporting of open activities and the accrual of credit both easy and normative. Some of the breakout groups explored the role the roundtable could play in promoting an easy, standard way for researchers to report open activities and incorporate open science into their workflow. Ashley Farley of the Bill & Melinda Gates Foundation noted that there is a need to measure the impact of open science and collaboration so that these activities might be properly rewarded. Agreement on open mandates, including dedicated funding, could accelerate this effort. Employers could also reward those who participate in open practices, demonstrating the value for all stakeholders and normalizing it as a practice.

Franko added that the roundtable could highlight the benefits of open behaviors, providing evidence against the "perceived" negatives. Standardizing the language for open practices and incentives could also support their broader use, said Konforti.

Jerry Sheehan of the National Institutes of Health raised the need for new approaches to accommodate a broader set of research outputs, includ-

including negative results and metrics, into assessments. He added that policies to shorten embargo periods and promote artificial intelligence-ready datasets could serve as motivators for data sharing and access, removing these barriers.

Katie Steen of the Association of American Universities advocated for further developing an infrastructure to support open scholarship systems. This could include, for example, conducting mapping and scoping exercises with an eye toward implementation and developing tools and methods for citing various research outputs that translate to metrics.

Adam Jones of the Gordon and Betty Moore Foundation discussed the need for credit and tools to support the use of open science related to personal persistent identifiers, capturing data downloads, and dataset usage.

(4) Build upon policies and processes already in place to support open science. Several breakout session participants encouraged the roundtable to lean into the building blocks that already exist in open science—for example, elevating success stories, supporting faculty champions, leveraging the role of libraries, and broadly disseminating toolkits. As Fitzpatrick and Farley noted, the roundtable can expand on existing efforts to share best practices to further promote open science.

Participants discussed the importance of identifying open science champions and disseminating success stories about their work. As Poldrack noted, the roundtable can work to establish more influential and prominent ways to highlight the work of open science champions, for example, through "Dean's Lists" and awards.

Heather Joseph of the Scholarly Publishing and Academic Resources Coalition proposed ways the roundtable can further build a coalition of the willing by challenging funders to create new programs to underwrite a diverse set of institutions to run open institution pilots. This could be supported with a network of university mentors to socialize successes.

Some participants emphasized the role of libraries in advocating for open science. Poldrack said the roundtable could raise the profile of library resources and share information about the potential for collaboration to support and mentor open science practices.

Hanisch recommended making success stories more visible, including those that address major challenges for recognition of open science and data sharing, especially discipline-dependent examples both within and outside the physical sciences. Franko also discussed the need to demonstrate the impact of open science with articles that showcase how expanding access to research can broaden knowledge.

Participants reflected on the potential of working with professional societies to help change norms in open science. Anderson suggested the roundtable systematically identify people who are willing and able to serve as champions for open research and scholarship in different professional societies. Broadly disseminating promising practices that have been adopted by professional societies, such as the Linguistic Society of America, American Geophysical Union, and Council of Graduate Departments of Psychology, among others, could encourage others to follow suit. Bourg agreed, adding that the roundtable should work closely with societies to move to open and overcome real and perceived barriers in this area. This could include, for example, partnering with societies to help their members share other kinds of open science research and outputs such as data, code, and publications.

Several participants discussed broader dissemination of the roundtable's toolkit and other resources to promote open science. For example, Konforti said that the roundtable should highlight how the toolkit is being leveraged at other institutions to incentivize change at universities that are not yet on board. Franko noted that the "Sending Signals Rubric" section of the toolkit can be used to intentionally engage university departments in an assessment of their level of engagement with open science. It can also be used to encourage institutional leadership to articulate where they stand on open science practices and address gaps between policy and on-the-ground practices at the institutional level. The tool could serve as a scoring system to guide institutions in a move toward open science practices.

Participants described the need to develop other tools and templates for promoting open science. Sheehan noted that the roundtable should support and provide guidance, models, and examples about open science practices including the use of repositories. Joseph discussed the need for template language for faculty senate resolutions and statements on institutional support for open science as well as fact sheets and data on the benefits of open science.

(5) Identify and address the financial and human costs of open science. Many participants commented that the roundtable should help to identify and evaluate the financial and human costs of open science honestly and proactively. To be successful in changing the culture, there should be an acknowledgment of the real costs of open science, who is burdened by those costs, and conversation about what action can be taken to address these costs at the university and discipline levels. Brooks Hanson of the American Geophysical Union and others discussed the importance of investing in

capacity, infrastructure, and people to ensure that open science is done well. Dougherty highlighted another specific cost—data archiving—as a major issue, especially in fields with large datasets. There is also a need for concrete financial support for open leaders through salaries, additional capacity, and funding, as stated by Gibson and other participants.

However, as Konforti noted, despite these initial costs, the larger benefits of open science, such as contributing to research on social justice issues and accelerating scientific discovery, should be communicated more broadly.

References

Lowndes, J. S. 2019. "Open Software Means Kinder Science." Scientific American *Observations* (blog), posted December 10, 2019. Available at https://blogs.scientificamerican.com/observations/open-software-means-kinder-science.

Lowndes, J. S., B. D. Best, C. Scarborough, J. C. Afflerbach, M. R. Frazier, C. C. O'Hara, N. Jiang, and B. S. Halpern. 2017. Our path to better science in less time using open data science tools. *Nature Ecology & Evolution* 1(0160). Available at https://www.nature.com/articles/s41559-017-0160.

NASEM (National Academies of Sciences, Engineering, and Medicine). 2018. *Open Science by Design: Realizing a Vision for 21st Century Research*. Washington, DC: The National Academies Press. https://doi.org/10.17226/25116.

Appendix A

Workshop Agenda

Developing a Toolkit for Fostering Open Science Practices: A Workshop

Roundtable on Aligning Incentives for Open Science

November 5, 2020

11:00 AM	**Opening Remarks and Goals for the Workshop from Roundtable Co-Chairs and Planning Committee Chair** Keith Yamamoto, University of California, San Francisco Loretta Parham, Atlanta University Center Tom Kalil, Schmidt Futures
11:15 AM	**Adopting and Utilizing a Toolkit for Open Science: Stakeholder Perspectives University Perspectives** *Thought Leaders*: Michael Crow, President, Arizona State University; and Philip E. Bourne, Founding Dean, University of Virginia School of Data Science *Researchers*: Tatiana Bryant, Research Librarian for Digital Humanities, History, and African American Studies, University of California, Irvine; and Camille Thomas, Technology and Digital Scholarship Scholarly Communications Librarian, Florida State University

Disciplinary Perspectives
Thought Leader: Lauren Collister, Chair, Committee on Scholarly Communication in Linguistics, Linguistic Society of America
Researcher: Sanjay Srivastava, Professor and Undergraduate Education Chair, Department of Psychology, University of Oregon

Research Funder Perspectives
Thought Leader: Ekemini Riley, Managing Director, Aligning Science Across Parkinson's
Researcher: Julia Stewart Lowndes, Senior Fellow, National Center for Ecological Analysis and Synthesis (NCEAS), University of California, Santa Barbara, and Founder, Openscapes

12:35 PM	**Break and Log-in to Breakout Sessions**
12:45 PM	**Stakeholder Breakout Sessions**
1:30 PM	**Breakout Reports and Concluding Discussion**
2:00 PM	**Adjourn Public Workshop**

Appendix B

Biographies of Speakers and Moderators

LORETTA PARHAM (*Workshop Chair*) is chief executive officer and library director of the Robert W. Woodruff Library of the Atlanta University Center, an independent entity operating as the single library shared by its four member institutions—Clark Atlanta University, the Interdenominational Theological Center, Morehouse College, and Spelman College. She is responsible for the strategic agenda transforming the Woodruff Library into the best choice for information for its community. With more than 30 years in the profession, her experience includes director of the Hampton University Library; deputy director of the Carnegie Library of Pittsburgh, Pennsylvania; district chief of the Chicago Public Library; and other public service positions with the Chicago Public Schools and the City Colleges of Chicago. An active leader, scholar, and engaging speaker, Ms. Parham was named a 2004 "Mover & Shaker" by Library Journal and was also honored as the 2017 Academic/Research Librarian of the Year by the Association of College and Research Libraries (ACRL). She has authored articles on Historically Black Colleges and Universities (HBCU) libraries and archives, and is co-editor of the book *Achieving Diversity: A How-To-Do-It Manual for Librarians*. Ms. Parham is co-founder and past chair of the HBCU Library Alliance, past chair of the Georgia Humanities Council, former board member of ACRL, the Wayne State University School of Library Science Advisory Board, and past treasurer of the Oberlin Group. Ms. Parham holds a master of library science degree from the University of Michigan, Ann Arbor, and a bachelor of science in communications from Southern Illinois

University, Carbondale. She is currently serving on the Board of Directors of EDUCAUSE and EDUCOPIA.

THOMAS KALIL (*Roundtable Co-Chair*) is chief innovation officer at Schmidt Futures. In this role, Mr. Kalil leads initiatives to harness technology for societal challenges, improve science policy, and identify and pursue 21st-century moonshots. Prior to Schmidt Futures, Mr. Kalil served in the White House for two presidents (Obama and Clinton), helping to design and launch national science and technology initiatives in areas such as nanotechnology, the BRAIN initiative, data science, materials by design, robotics, commercial space, high-speed networks, access to capital for startups, high-skill immigration, STEM education, learning technology, startup ecosystems, and the federal use of incentive prizes. From 2001 to 2008, Mr. Kalil was special assistant to the chancellor for science and technology at the University of California, Berkeley. Mr. Kalil received a B.A. in political science and international economics from the University of Wisconsin–Madison and completed graduate work at the Fletcher School of Law and Diplomacy.

KEITH R. YAMAMOTO (NAS/NAM) (*Roundtable Co-Chair*) is University of California, San Francisco (UCSF) vice chancellor for science policy and strategy, director of precision medicine for UCSF, and professor of cellular and molecular pharmacology at UCSF. He is a leading researcher investigating transcriptional regulation by nuclear receptors, which mediate the actions of essential hormones and cellular signals; he uses mechanistic and systems approaches to pursue these problems in pure molecules, cells, and whole organisms. He has led or served on numerous national committees focused on public and scientific policy, public understanding and support of biological research, and science education; he chairs the Coalition for the Life Sciences, and sits on the National Academy of Medicine Council and the National Academies Division of Earth and Life Studies Advisory Committee. As chair of the Board on Life Sciences, he created the study committee that produced *Toward Precision Medicine: Building a Knowledge Network for Biomedical Research and a New Taxonomy of Disease*, the report that enunciated the precision medicine concept, and he has helped to lead efforts in the White House, in Congress, in Sacramento, and at UCSF to implement it. He has chaired or served on many committees that oversee training and the biomedical workforce, research funding, and the process of peer review and the policies that govern it at the National Institutes of Health. He is a member of the advisory board for Lawrence Berkeley National Laboratory and the board of directors of Research!America. He was elected to the National Academy of Sciences, the National Academy of Medicine, the American Academy of Arts and Sciences, and the American Academy of Microbiology, and is a fellow of the American Association for the Advancement of Science.

PHILIP E. BOURNE is the founding dean of the School of Data Science and professor of biomedical engineering at the University of Virginia (UVA). From 2014 to 2017, Dr. Bourne was the associate director for data science at the National Institutes of Health (NIH). In this role he led the Big Data to Knowledge Program, coordinating access to and analyzing biomedical research from across the globe and making it available to scientists and researchers. While there, he was also responsible for governance and strategic planning activities for data and knowledge management, and established multiple trainings in data science. He has done exceptional work to make biomedical research accessible, as well as to advance the field of data science. Prior to his time at the NIH, Dr. Bourne spent 20 years on the faculty at the University of California, San Diego, eventually becoming associate vice chancellor of innovation and industrial alliances. He is a highly respected and oft-cited scholar who brings a wealth of experience to UVA.

TATIANA BRYANT is the research librarian for digital humanities, history, and African American studies at the University of California, Irvine, libraries. She holds an M.P.A. in international public and nonprofit administration, management, and policy from New York University; an M.S. in information and library science from Pratt Institute; and a B.A. in history from Hampton University. She has been a SPARC OpenCon Berlin fellow and a Digital Native American and Indigenous Studies fellow through the National Endowment for the Humanities Office of Digital Humanities Institute. Her research includes studies on gender identity and performance in library work as well as perceptions of open access publishing among faculty who identify as Black, Indigenous, and/or people of color. She is a co-editor of the forthcoming volume *Implementing Excellence in Diversity, Equity, and Inclusion: A Handbook for Academic Libraries* (ACRL Press).

MICHAEL CROW became the 16th president of Arizona State University (ASU) on July 1, 2002. He is guiding the transformation of ASU into one of the nation's leading public metropolitan research universities, an institution that combines the highest levels of academic excellence, inclusiveness to a broad demographic, and maximum societal impact—a model he terms the "New American University." During his tenure, the university more than quadrupled research expenditures, completed an unprecedented infrastructure expansion, and was named the nation's most innovative school by *U.S. News & World Report* in 2016, 2017, and 2018. Dr. Crow was previously executive vice provost of Columbia University, where he also was professor of science and technology policy in the School of International and Public Affairs. He played the lead role in the creation of, and served as the founding director of, the Earth Institute at Columbia University, and in 1998 founded the Consortium for Science, Policy, and Outcomes, dedicated to linking science and technology to optimal social, economic, and environmental outcomes. An elected fellow of the American Association for the Advancement of Science and the National Academy of Public Administration, and member of the Council on Foreign Relations and U.S. Department of Commerce National Advisory Council on Innovation and Entrepreneurship, he is the author of books and articles analyzing science and technology policy and the design of knowledge enterprises and higher education institutions and systems. Dr. Crow received his Ph.D. in public administration (science and technology policy) from the Maxwell School of Citizenship and Public Affairs, Syracuse University.

LAUREN COLLISTER is chair of the Committee on Scholarly Communication in Linguistics at the Linguistic Society of America. Dr. Collister is also the director of the Office of Scholarly Communication and Publishing at the University Library System, University of Pittsburgh. She oversees all of the university's open access publishing, repository, copyright, altmetrics, and scholarly communication work. Broadly, she is an advocate for open scholarship, and her work is to advance the open scholarship movement. Her research interests in linguistics include discourse markers and deixis and how those intersect with linguistic innovation in online spaces. She is a member of the Linguistics Data Interest Group for the Research Data Alliance and also works as an advocate for good scholarly communication practice within the field of linguistics. Dr. Collister has a Ph.D. in sociolinguistics and has since worked to help linguists share their research openly, including advancing the data citation and sharing practices of that field.

JULIA STEWART LOWNDES is senior fellow at the National Center for Ecological Analysis and Synthesis (NCEAS) of the University of California, Santa Barbara, and founding director of Openscapes. Dr. Lowndes champions kinder, better science in less time through open data science and teamwork. As a marine data scientist, 2019 Mozilla fellow, and senior fellow at NCEAS, she has more than 7 years designing and leading programs to empower science teams with skillsets and mindsets for reproducible research, empowering researchers with existing open tools and communities. She has been building communities of practice in this space since 2013 with the Ocean Health Index, after earning her Ph.D. at Stanford University studying drivers and impacts of Humboldt squid in a changing climate. Dr. Lowndes is a Carpentries instructor, lead creator of the Ocean Health Index's open data science training, and a co-founder of Eco-Data-Science and R-Ladies Santa Barbara.

EKEMINI RILEY is the managing director of Aligning Science Across Parkinson's (ASAP), a research-funding initiative that coordinates targeted basic research and resources to uncover the roots of Parkinson's disease. Prior to ASAP, Dr. Riley was a director at the Milken Institute Center for Strategic Philanthropy where she helped shape and co-direct their medical research practice. She designed and facilitated several multisector think-tank sessions to inform the strategic deployment of philanthropic capital, crafted research programs, and seeded multifunder collaboration. Dr. Riley led the development and launch of ASAP, as well as the Gilbert Family Foundation's Gene Therapy and Vision Restoration Initiatives. Her work also laid the foundation for Play It Forward Pittsburgh, an organ donation awareness campaign. Dr. Riley completed her B.A. in natural sciences from the Johns Hopkins University and Ph.D. in molecular medicine from the University of Maryland School of Medicine. Her doctoral research focused on gene regulation of an endogenous protease inhibitor and its role in innate immunity and tumor suppression.

SANJAY SRIVASTAVA is a professor and undergraduate education chair in the Department of Psychology at the University of Oregon and director of the Personality and Social Dynamics Lab. He teaches courses on several topics, including introductory psychology, motivation and emotion, social and personality psychology, and advanced statistics. Prior to coming to the University of Oregon, Dr. Srivastava was a postdoctoral research scientist at Stanford University. His research focuses on how personality affects and is affected by the social environment. This includes research on interpersonal perception, emotions, personality dynamics and development, and the psychology of online societies. He received his B.A. in psychology from Northwestern University and his Ph.D. in psychology from the University of California, Berkeley.

CAMILLE THOMAS is the scholarly communications librarian at Florida State University. She currently leads initiatives to support students, faculty, and staff to engage with new modes of research and teaching, including open access and open education. She has worked as a SPARC fellow, on public interest partners and enhancing discovery for open educational resources. Ms. Thomas received her master's degree in library and information studies from Florida State University in 2015 and a B.A. in creative writing and journalism from the University of Central Florida in 2012. Her research includes data in libraries, early-career leadership and management, user experience and open access, and open education.

Appendix C

Toolkit Elements

This appendix includes examples of draft elements of a toolkit that have been developed by members of working groups of the National Academies of Sciences, Engineering, and Medicine's Roundtable on Aligning Incentives for Open Science. The following materials were developed to stimulate discussions at the November 5, 2020, workshop on Developing a Toolkit for Fostering Open Science Practices:

I. **Open Science Imperative**. This essay communicates the benefits of open science using approachable language.
II. **Open Science Signaling Language Template and Rubric**. These resources provide specific language that can be adapted and adopted to signal an organization's interest in open science activities at specific points of high leverage (e.g., grant applications, job postings).
III. **Good Practices Primers**. These concise guides offer policy makers a high-level overview of open sharing.
IV. **Open Science by the Numbers Infographic**. This infographic communicates the benefits of open science in a graphic form.
V. **Open Science Success Stories Database**. This database compiles research articles, perspectives, case studies, news stories, and other materials that demonstrate the myriad ways in which open science benefits researchers and society alike.

VI. **Reimagining Outputs Worksheet**. This table enumerates the range of research products stakeholders may choose to consider as they develop open science policies.

The toolkit is primarily intended to assist university leadership, academic department chairs, research funders, learned societies, and government agencies about how such a toolkit might be used, what additional materials are needed, and how such a toolkit should be disseminated for broad adoption. As a result of the workshop, a few sections in the Open Science Imperative and Good Practices Primers have been revised by the working group authors.

I. OPEN SCIENCE IMPERATIVE[1]

Derrick Anderson, Arizona State University
Rachel Bruce, UK Research and Innovation
Ashley Farley, Bill & Melinda Gates Foundation
Robert Hanisch, National Institute of Standards and Technology
Greg Tananbaum, Open Research Funders Group
Thomas Wang, American Heart Association/
University of Texas Southwestern Medical Center

This narrative communicates the benefits of open science using succinct, approachable language. One way to think about its possible deployment is to envision an academic administrator or senior leader at a philanthropy who has a vague notion that open science is something they should better understand. This piece, if successfully executed, will make the affirmative case as to why the open approach to the research endeavor is preferable to the status quo, and what the benefits to society will be if it is adopted at scale.

Over the last 20 years, the research community has grown increasingly interested in and supportive of open science activities. Open science encompasses a range of individual, institutional, and community efforts to broaden access to research outputs. This increased accessibility facilitates better collaboration and outcomes as a function of collective intelligence. By prioritizing shared discovery over individual and institutional agendas, open science practices are spurring the knowledge economy, generating broad social and public benefits, strengthening cultural values for scientific literacy and education, and improving public policy and democracy (Tennant et al., 2016; Zuccala, 2010). Despite the benefits of open science, individual researchers face numerous barriers that are restricting broad uptake of these practices. The current credit and reward systems disincentivize information sharing in favor of siloed, noninclusive modes of knowledge production. Significant, coordinated support within and across research stakeholder groups is necessary to change these incentives to realize the benefits of open science. This white paper, prepared in conjunction with the National Academies of Sciences, Engineering, and Medicine's Roundtable on Align-

---

[1] The views expressed are those of the authors and do not necessarily reflect the official policies or positions of their employing organizations.

ing Incentives for Open Science, briefly sketches the current state of open science, contrasts the diminishing returns of the traditional scientific model with the advantages of emergent open science practices, and suggests possible measures that organizations can individually and collectively undertake to shape the future of research and discovery.

THE STATE OF OPEN SCIENCE

Open science has been conceptualized in philosophical and ideological terms as an affinity for open flows of information to facilitate innovation for the betterment of society (Gold, 2016), but it is most frequently used as an umbrella term to describe active efforts to reduce the barriers to information access for researchers and the public. A commonly used definition of open science is "the idea that scientific knowledge of all kinds should be openly shared as early as is practical in the discovery process" (Nielsen, 2011). Although varying conceptualizations and definitions of open science exist, there is general agreement on the practices that support it, such as open access publication, research preregistration, open access to data and materials, and development of open source software (Berg and Niemeyer, 2018; Gold, 2016; Gold et al., 2019).

Increased adoption of these mutually reinforcing practices by institutions and especially by individual researchers has created a momentum behind open science. This momentum is reflected partly by the choices that researchers make regarding how their data are shared. In one survey, the number of researchers who reported making their data openly available increased from just over 55 percent to 64 percent between 2016 and 2018. From before 1990 through the 2010s, the percentage of researchers who were unaware of the license under which they made their data openly available decreased from 71 percent to 54 percent. During the same time, the percentage of respondents who would feel motivated to make their data openly available for co-author credit increased from 7 percent to 27 percent (Digital Science and Figshare, 2018, 8, 13).

The rise of open access as a widespread publishing practice also indicates greater uptake of open science principles and values. An analysis of 70 million articles published between 1950 and 2019 determined that at least 31 percent of all scholarly publications are available as open access and that the proportion is growing. The same analysis indicated that, given existing trends, 70 percent of all article views will be to open access papers by 2025 (Piwowar et al., 2019). This trend appears to be driven by the values held by researchers: "Over 90 percent of OA [open access] authors published this way because of the principle of free access" (Swan and Brown, 2004, 5) and because of "their perceptions that these journals reach larger audiences, publish more rapidly and are more prestigious than the toll-access (subscription-based) journals that they have traditionally published in" (Swan and Brown, 2005, ES 1). This momentum toward the open sharing of research papers is further underscored by the spectacular flourishing of preprints, with both readership and authorship growth near 100 percent year-on-year (Abdill and Blekhman, 2019).

These data indicate that although open science practices have been adopted by an increasing number of researchers, a large share of researchers remain either unaware of the benefits of these practices or find that the barriers to adoption (including time, resources, lack of clear guidance, and ambiguous incentives) are significant. Enhanced researcher awareness and adoption of open science approaches, combined with proper institutional support and better alignment of credit/reward systems, holds the potential to realize greater knowledge diffusion; improved efficiency, transparency, and interdisciplinarity of scientific exploration; and a more robust, accessible, and replicable body of research (Spellman et al., 2018; Tennant et al., 2016).

THE BENEFITS OF OPEN SCIENCE

Communicating the advantages of open science to researchers and the broader public is essential to greater uptake of these practices. Open science offers an array of benefits across five domains:

1. *Supporting the growth of the knowledge economy.* By facilitating freer flows of information among scientists, research institutions, and firms, open science practices can accelerate the discovery process and commercialization of scientific research. The inherently transparent nature of open science also makes testing the reproducibility and replicability of scientific research substantially more efficient.
2. *Improving the integrity, reliability and transparency of scientific research.* Science as a process operates with reproducibility as a core objective. Students are trained through replication exercises and scientists are expected to describe their work in ways that facilitate replication. Open science practices make the processes of

science more transparent, which, in turn, makes scientific findings easier to test and to trust.
3. ***Generating social and public benefit.*** By lowering barriers to public participation in science, open science approaches allow social needs articulated by the public to inform a greater share of scientific research and enable citizens to make better-informed decisions.
4. ***Strengthening scientific literacy and education.*** By making scientific research freely available to the public, open science enables nonscientists to become more familiar with scientific methods and encourages greater layperson interest in applying a rigorous, inquisitive approach to their engagement with the world and the pressing issues of the day.
5. ***Improving public policy and democracy.*** By encouraging greater transparency in research and availability of research products, open science allows policy makers and the public to be more informed about research that can be used to shape policy and promote civic action.

Numerous research projects and platforms have realized the benefits of open science approaches, sometimes across all five of these domains, including the following:

- **The Human Genome Project,** completed in 2003, was carried out with an explicit commitment to open science. Participating researchers pledged to make their discoveries available online within 24 hours and provide unrestricted access to information in real time. As a result, the project's public-domain gene sequences generated an estimated 30 percent more genetic diagnostic tests than genes that were first sequenced by private firms and then restricted as intellectual property. The myriad of public and private economic benefits created by the Human Genome Project (estimated at $965 billion and nearly four million jobs between 1988 and 2012; Tripp and Grueber, 2011) have established it as a model for the effective use of open data, providing a picture of what the future of science and innovation could look like with greater adoption of open science practices (SPARC, n.d.).
- The **Group on Earth Observations (GEO)** is a global network of more than 100 national governments and more than 100 par-

ticipating organizations that enables the collection and sharing of atmospheric, oceanic, and terrestrial data and information to facilitate better decision making and policy formulation. GEO's Global Earth Observation System of Systems (GEOSS) portal was designed according to best practices in open science to facilitate open, coordinated, and sustained data sharing to advance the United Nations 2030 Agenda for Sustainable Development, the Paris Agreement, and the Sendai Framework for Disaster Risk Reduction. In addition to enabling communication between researchers and governments, "data products and information derived from GEO data can be useful for individuals to better understand the environment in which they live and work, and protecting the health of their family, and better educating themselves, and through the positive results of many other generative and even serendipitous applications" (Benkler, 2006; Mayo and Steinberg, 2007; NRC, 2009; and Zittrain, 2006; cited in Uhlir, 2015, 13).

- **The Lab @ DC** is a unit within the Washington, D.C., mayor's administration that works to design public policy and program interventions for the city. The Lab @ DC uses the Open Science Framework to share their methodology, analysis, and evaluations of municipal programs, utilizing transparency to allow their projects to be reproduced and replicated by other community groups. Projects that have been undertaken by this group span from transit, housing, and public safety to customer service and economic prosperity (The Lab @ DC, n.d.).

- **Symbiota** is an exclusively web-based open source content management system that integrates natural history collections and other biological community knowledge and data into a network of databases and tools to increase knowledge of biodiversity. Since 2012, 73 percent of projects funded by the National Science Foundation's Advancing Digitization of Biodiversity Collections have used Symbiota. The platform now hosts 37 million records from 766 universities, museums, and research organizations, including linkages to images, tissues, DNA sequences, and taxonomic and ecological information (Symbiota, n.d.). Importantly, Symbiota's software design philosophy and implementation was driven by its "*user community*—e.g., collections managers, taxonomists, ecologists, data entry personnel, programmers, informaticians, and students" (Gries et al., 2014). Symbiota is freely available to researchers and the public.

- **Global Open Data for Agriculture and Nutrition (GODAN)** is an initiative of the U.S. Department of Agriculture and U.S. Agency for International Development that promotes open data sharing to increase global access to information about agriculture and nutrition. Leveraging data input from a partner network of more than 700 private- and public-sector, nonprofit, and academic organizations, GODAN aims to inform and improve daily decision making for farmers and consumers, with the goal of developing solutions to global hunger (Adams, 2018).
- **Microreact** is a free, real-time tool for visualizing and tracking outbreaks of diseases such as Ebola and Zika, as well as antibiotic-resistant microbes. Developed through a collaboration between researchers from the Wellcome Trust Sanger Institute and Imperial College London, Microreact allows any researcher in the world to upload information on disease outbreaks via its web browser, which can be shared and visualized through Microreact's cloud-based system. Microreact also integrates data submitted for publication in the journal *Microbial Genomics* to encourage greater data availability and access (Wellcome Trust, 2016).
- The **California Policy Lab** is a nonprofit based at the University of California, Los Angeles and Berkeley, that partners with state and local governments to solve social issues, including homelessness, poverty, crime, and educational inequality (California Policy Lab, n.d.). The California Policy Lab utilizes the Open Science Framework and has established data-sharing agreements with more than a dozen county agencies in Los Angeles, Sonoma, and San Francisco covering "medical, mental health, criminal justice, social service, and homeless management information systems" (California Policy Lab, 2018). The lab recently received a $1.2 million grant to expand to all University of California schools and partner with more public agencies to conduct policy-relevant research and overcome data silos.
- The **International Virtual Observatory Alliance** is an open platform enabling astronomers, educators, and the general public to discover, access, and integrate open data from worldwide (including in orbit) observatories. It links together the vast astronomical archives and databases around the world, together with analysis tools and computational services, into a single, integrated

facility. From its inception in 2002 through late 2020, the Virtual Observatory data have powered more than 2,300 scholarly papers,[2] covering the entire electromagnetic spectrum, from gamma-rays to radio waves.
- The **COVID-19 Open Research Dataset (CORD-19)** is an open collection of scientific articles and preprints related to COVID-19 and historical coronavirus research. The dataset can be text mined and analyzed using artificial intelligence and natural language processing to generate new insights into combatting the virus. The dataset was downloaded more than 200,000 times in the first 3 months after its release (Wang et al., 2020). This is one of several examples of open science's centrality in rapidly addressing this era's most critical public health challenge.

OPEN SCIENCE AND THE STATUS QUO

Historically, academic research environments have incentivized competition between individual researchers, which stymies collaboration and leads to the hoarding of knowledge. These dynamics persist as a function of the pursuit of "excellence" by research institutions, which results in the widespread usage of metrics that decrease transparency and collaboration. For example, measuring success by the number of patents filed and industry spinoffs launched leads to the safeguarding of intellectual property by researchers rather than the sharing of this information with external organizations that can increase the possibility of taking a product to market. Likewise, when academic departments measure their success by the volume of research citations and grant tenure to researchers who are cited most frequently, researchers are pressured to be the first to publish their findings and often operate in isolation, rarely venturing out of their respective research programs and communities (Heenan and Williams, 2018). Researchers become understandably hesitant to make their data and findings openly available out of fear of being "scooped" by other researchers (Berg and Niemeyer, 2018). Although competition between institutions and individual researchers may have been adequate to drive discovery in the 20th century, the "explosive sophistication" of science and engineering fields, in particular, has made it impossible for a single individual to be an

---

[2] Data accessed from the SAO/NASA Astrophysics Data System, October 16, 2020.

expert in multiple specialties or even a single subfield. Effective knowledge production now demands teams of researchers with diverse knowledge and skills to facilitate ongoing discovery (Brooks, 2010). Greater collaboration, rather than being an aspirational ideal that might produce better outcomes under the right circumstances, has now become a necessity to contend with the extreme specialization of knowledge production and ensure that discovery continues apace.

Open science practices, in contrast to traditional models of knowledge production, emphasize that open, transparent, and collaborative research dissemination practices more properly balance collective, institutional, and individual benefits. Open science represents a positive evolution of the research endeavor along three dimensions:

- ***Collaboration drives innovation with the potential for broad social impact.*** Open science approaches can reduce barriers between researchers and other stakeholders, including the public (e.g., by better informing and directly involving patients in biosciences) (Gold, 2016). By making data openly accessible between researchers and the public, open science can provide greater opportunities for interdisciplinary, collaborative research across institutions worldwide (Uhlir, 2015). Heightened collaboration can also lead to dynamic new knowledge hubs and remove barriers to upstream research and technology transfer (Gold, 2016).
- ***Greater efficiency and speed.*** Open data practices also drive efficiency by enabling real-time, data-driven decision making (Adams, 2018; SPARC, n.d.). The sharing of data reduces transaction costs; increases reproducibility and reuse of data; decreases redundancy; and drives greater transparency, heightened efficiency, and accelerated sustainable innovation (Gold, 2016; Gold et al., 2019; Tennant et al., 2016).
- ***Replicability enhances trust and research quality.*** By enhancing researchers' ability to verify results, open science practices help to build trust and goodwill among researchers and enhance the legitimacy of research (Popkin, 2019; Uhlir, 2015).

THE ROLE OF RESEARCH STAKEHOLDER ORGANIZATIONS

Open science has been largely pioneered by individual researchers who believe the benefits of this approach—to their work, to the shared understanding of a problem space, to their discipline, and to society—outweigh the reputational benefits that may be derived from the older, competition-based models of knowledge production. However, many researchers continue to face strong disincentives for engaging in open science practices, especially early-career scholars, who face the greatest pressure to conform to the traditional modes of credit and recognition that can lead to tenure. The wider uptake of open science, therefore, requires the organizational stakeholders responsible for reward systems—institutions, government agencies, and philanthropies chief among them—to establish new incentives and processes that prioritize open science activities. Because the competition-based incentives that motivate researchers reflect institutional prerogatives to demonstrate excellence vis-à-vis other institutions, institutions must also convene to identify new approaches toward facilitating interinstitutional collaboration and collectively address external barriers to open science.

Fortunately, the values that underpin open science—such as inclusiveness, collaboration, social impact, and scientific literacy—are mutually reinforcing to the missions of the research institutions, agencies, and funding organizations that support scientific research. Forward-thinking organizations have already begun to implement incentives for open science practices that provide a model for others to follow, which have taken several forms, including the following:

1. ***Creating supportive environments.*** The Tanenbaum Open Science Institute (TOSI) at the Neuro (Montreal Neurological Institute-Hospital) was designed as a "living lab for Open Science" to achieve the goals of accelerating discovery in neuroscience through collaboration, developing global best practices, and delivering innovative treatment to benefit patients afflicted by neurological diseases. TOSI supports four Open Science initiatives, including a biologic imaging and genetic repository, an open research platform, several open neuroinformatics platforms, and an early-stage drug discovery unit that collaborates with academia and industry partners (Gold, 2016; Neuro, n.d.).

2. ***Incentivizing open access publishing.*** The Bill & Melinda Gates Foundation and the Wellcome Trust, which funded $1.3 billion and $1.2 billion in global health research, respectively, joined a consortium of 11 European funding agencies that require all funded research to be free immediately upon publication. This incentive effectively requires scientists to publish papers in open access journals rather than those that charge subscriptions (Stokstad, 2018).
3. ***Awards for Open Science innovation.*** In 2017 the National Institutes of Health, Wellcome Trust, and the Howard Hughes Medical Institute hosted the Open Science Prize competition, leveraging public input to determine award finalists (NIH, 2017).

These examples represent the kinds of new incentives critical to instantiating the cultural shift necessary for sustained uptake of Open Science. In designing new incentives, research organizations and funders may also consider topics such as advancing the theory and practice of Open Science; how hiring decisions may contribute to cultures supportive of Open Science; and how funding mechanisms can be evolved to encourage open access publishing, data archiving and sharing, preregistration, and collaboration. The National Academies' Roundtable on Aligning Incentives for Open Science aims to encourage exploration of these topics and a wide range of possibilities for using incentives to realize the full potential for scientific research as a catalyst for discovery, economic growth, and societal benefit.

REFERENCES

Abdill, R. J., and R. Blekhman. 2019. Meta-Research: Tracking the popularity and outcomes of all bioRxiv preprints. *eLife* 8:e45133. DOI: 10.7554/eLife.45133.

Adams, J. 2018. Open Data: Enabling Fact-Based, Data-Driven Decisions. U.S. Department of Agriculture (blog). Available at https://www.usda.gov/media/blog/2018/07/13/open-data-enabling-fact-based-data-driven-decisions. Accessed January 19, 2021.

Benkler, Y. 2006. The Wealth of Networks—How Social Production Transforms Markets and Freedom. Yale University Press. Available at http://www.benkler.org/Benkler_Wealth_Of_Networks.pdf. Accessed August 30, 2021.

Berg, D. R., and K. E. Niemeyer. 2018. The case for openness in engineering research. F1000Research 7:501.

Brooks, Jr., F. P. 2010. *The Design of Design: Essays from a Computer Scientist.* London: Pearson Education.
California Policy Lab. 2018. California Policy Lab Awarded $1.2M UC Multicampus Research Grant. Press release, December 13, 2018. Available at https://www.capolicylab.org/wp-content/uploads/2018/12/CPL-Press-Release-re-MRPI-12-12-18-final.pdf. Accessed January 19, 2021.
California Policy Lab. n.d. What we do. Available at https://www.capolicylab.org/what-we-do. Accessed January 19, 2021.
Digital Science and Figshare. 2018. *The State of Open Data Report 2018.* Available at https://figshare.com/articles/report/The_State_of_Open_Data_Report_2018/7195058. Accessed January 19, 2021.
Gold, E. R. 2016. Accelerating translational research through open science: The Neuro Experiment. *PLOS Biology* 14(12):e2001259.
Gold, E. R., S. E. Ali-Khan, L. Allen, L. Ballell, M. Barral-Netto, D. Carr, D. Chalaud, S. Chaplin, M. S. Clancy, P. Clarke, R. Cook-Deegan, A. P. Dinsmore, M. Doerr, L. Federer, S. A. Hill, N. Jacobs, A. Jean, O. A. Jefferson, C. Jones, L. J. Kahl, T. M. Kariuk, S. N. Kassell, R. Kiley, E. R. Kittrie, B. Kramer, W. H. Lee, E. MacDonald, L. M. Mangravite, E. Marincola, D. Mietchen, J. C. Molloy, M. Namchuk, B. A. Nosek, S. Paquet, C. Pirmez, A. Seyller, M. Skingle, S. N. Spadotto, S. Staniszewska, and M. Thelwall. 2019. An open toolkit for tracking open science partnership implementation and impact. *Gates Open Research* 3:1442. Available at https://doi.org/10.12688/gatesopenres.12958.1. Accessed January 19, 2021.
Gries, C., E. Gilbert, and N. Franz. 2014. Symbiota – A virtual platform for creating voucher-based biodiversity information communities. *Biodiversity Data Journal* 2:e1114. DOI: 10.3897/BDJ.2.e1114.
Heenan, A., and I. D. Williams. 2018. How open data can help the world better manage coral reefs. *The Conversation* (January 29). Available at https://theconversation.com/how-open-data-can-help-the-world-better-manage-coral-reefs-88805. Accessed January 19, 2021.
Mayo, E., and T. Steinberg. 2007. The Power of Information: An Independent Review. Available at http://ses.library.usyd.edu.au/bitstream/2123/6557/1/PSI_vol2_chapter20.pdf. Accessed August 30, 2021.
Neuro. n.d. Open science, to accelerate discovery and deliver cures. Available at https://www.mcgill.ca/neuro/open-science-0. Accessed January 19, 2021.
Nielsen, M. 2011. An informal definition of open science. OpenScience Project (website). Available at http://openscience.org/an-informal-definition-of-openscience. Accessed January 19, 2021.
NIH (National Institutes of Health). 2017. Open Science Prize announces epidemic tracking tool as grand prize winner. Press release, February 28, 2017. Available at https://www.nih.gov/news-events/news-releases/open-science-prize-announces-epidemic-tracking-tool-grand-prize-winner. Accessed January 19, 2021.

NRC (National Research Council). 2009. *The Socioeconomic Effects of Public Sector Information on Digital Networks: Toward a Better Understanding of Different Access and Reuse Policies: Workshop Summary.* Washington, DC: The National Academies Press. https://doi.org/10.17226/12687.

Piwowar, H., J. Priem, and R. Orr. 2019. The Future of OA: A large-scale analysis projecting Open Access publication and readership. *bioRxiv* 795310. DOI: https://doi.org/10.1101/795310.

Popkin, G. 2019. Data sharing and how it can benefit your scientific career. *Nature* (Career Feature article), May 13, 2019. Available at https://www.nature.com/articles/d41586-019-01506-x. Accessed January 19, 2021.

SPARC (Scholarly Publishing and Academic Resources Coalition). n.d. "From ideas to industries: Human Genome Project." Available at https://sparcopen.org/impact-story/human-genome-project. Accessed January 19, 2021.

Spellman, B. A., E. A. Gilbert, and K. S. Corker. 2018. Open science. In *Stevens' Handbook of Experimental Psychology and Cognitive Neuroscience: Volume 5 Methodology*, 1–47. New York: John Wiley & Sons.

Stokstad, E. 2018. In a win for open access, two major funders won't cover publishing and hybrid journals. *Science.* DOI:10.1126/science.aav9422.

Swan, A., and S. Brown. 2004. Authors and open access publishing. *Learned Publishing* 17:219–224. DOI:10.1087/095315104323159649.

Swan, A., and S. Brown, S. 2005. Open access self-archiving: An author study. Key Perspectives (website). Available at http://cogprints.org/4385/1/jisc2.pdf. Accessed January 19, 2021.

Symbiota. n.d. Symbiota Introduction. Available at http://symbiota.org/docs. Accessed January 19, 2021.

Tennant, J. P., F. Waldner, D. C. Jacques, P. Masuzzo, L. B. Collister, and C. H. J. Hartgerink. 2016. The academic, economic and societal impacts of Open Access: An evidence-based review. F1000Research 5:632. Available at https://f1000research.com/articles/5-632. Accessed January 19, 2021.

The Lab @ DC. n.d. Home. Available at https://thelab.dc.gov. Accessed January 19, 2021.

Tripp, S., and M. Grueber. 2011. *Economic Impact of the Human Genome Project.* Battelle Memorial Institute, Technology Partnership Practice. Available at https://www.battelle.org/docs/default-source/misc/battelle-2011-misc-economic-impact-human-genome-project.pdf?sfvrsn=6. Accessed January 19, 2021.

Uhlir, P. 2015. The Value of Open Data Sharing – A White Paper for the Group on Earth Observations. Group on Earth Observations, GEO-XII Plenary and Mexico City Ministerial Summit, November 11–12, 2015. Available at https://www.earthobservations.org/documents/geo_xii/GEO-XII_09_The%20Value%20of%20Open%20Data%20Sharing.pdf.

Wang, L. L., K. Lo, Y. Chandrasekhar, R. Reas, J. Yang, D. Burdick, D. Eide, K. Funk, Y. Katsis, R. Kinney, Y. Li, Z. Liu, W. Merrill, P. Mooney, D. Murdick, D. Rishi, J. Sheehan, Z. Shen, B. Stilson, A. Wade, K. Wang, N. X. R. Wang, C. Wilhelm, B. Xie, D. Raymond, D. S. Weld, O. Etzioni, and S. Kohlmeier. 2020. CORD-19: The Covid-19 Open Research Dataset. *ArXiv* [preprint]. April 22, 2020. arXiv:2004.10706v2.

Wellcome Trust Sanger Institute. 2016. Online epidemic tracking tool embraces open data and collective intelligence to understand outbreaks. News article, November 30, 2016. Available at https://www.sanger.ac.uk/news/view/online-epidemic-tracking-tool-embraces-open-data-and-collective-intelligence-understand. Accessed January 19, 2021.

Zittrain, J. 2006. The Generative Internet. *Harvard Law Review* 119. Available at http://dash.harvard.edu/handle/1/9385626. Accessed August 30, 2021.

Zuccala, J. 2010. Open access and civic scientific information literacy. *Information Research: An International Electronic Journal* 15(1).

II. OPEN SCIENCE SIGNALING LANGUAGE TEMPLATE AND RUBRIC[3]

Maryrose Franko, Health Research Alliance
Courtney Brown, Lumina Foundation
Rachel Bruce, UK Research and Innovation
Glenn Dillon, American Heart Association
Randolph Hall, University of Southern California
Robert Kiley, Wellcome Trust
Lisa Nichols, Formerly, Office of Science and Technology Policy
Greg Tananbaum, Open Research Funders Group
Roger Wakimoto, University of California, Los Angeles

This resource provides specific language that can be adapted and adopted to signal an organization's interest in open science activities at specific points of high leverage (e.g., grant applications, job postings). Even absent adoption of formal open science policies, this language can indicate an organization's values and "nudge" researcher behavior toward open practices.

NOTE: The language below can be customized to reflect the specific research considerations of each participating organization.

FUNDERS AND AGENCIES

Grant Application

1. Foundation XYZ values the open sharing of research outputs. If applicable, describe (1) instances where you have engaged in "open" activities (such as making articles open access and sharing data/code according to FAIR principles [Findability, Accessibility, Interoperability, and Reuse of digital assets]); (2) examples of how your open research outputs have been used by others in your discipline, in other disciplines, and/or outside of academia (include DOIs if possible); and (3) plans to engage in open activities in the future.

---

[3] The views expressed are those of the authors and do not necessarily reflect the official policies or positions of their employing organizations.

APPENDIX C

2. For each of the categories below, provide *representative examples* demonstrating how you have made research outputs resulting from other projects openly accessible. If possible, provide the DOI and license terms under which the materials are available.

- Open access articles
- Open access books, book chapters, and/or monographs
- Copies of your papers, chapters, monographs, or other published materials in institutional or disciplinary repositories
- Preprints
- Datasets
- Software/Code
- Materials/Reagents
- Preregistration plans
- Other outputs (please describe)

Additionally, it is important to include negative and null results, which could be covered in a variety of information formats.

Grant Progress Report

1. Foundation XYZ values the open sharing of research outputs. If applicable, describe, in the context of this funded project, (1) instances where you have engaged in "open" activities (such as making articles open access and sharing data/code according to FAIR principles); (2) examples of how your open research outputs have been used by others in your discipline, in other disciplines, and/or outside of academia (include DOIs if possible); and (3) plans to engage in open activities as the project progresses and concludes.

2. For each of the categories below, provide *representative examples* demonstrating how you have made research outputs resulting from this project openly accessible. If possible, provide the DOI and license terms under which the materials are available.

- Open access articles
- Open access books, book chapters, and/or monographs

- Copies of your papers, chapters, monographs, or other published materials in institutional or disciplinary repositories
- Preprints
- Datasets
- Software/Code
- Materials/Reagents
- Preregistration plans
- Other outputs (please describe)

Additionally, it is important to include negative and null results, which could be covered in a variety of information formats.

Grant Final Report

1. Foundation XYZ values the open sharing of research outputs. If applicable, describe, in the context of this funded project, (1) instances where you have engaged in "open" activities (such as making articles open access and sharing data/code according to FAIR principles); (2) examples of how your open research outputs have been used by others in your discipline, in other disciplines, and/or outside of academia (include DOIs if possible); and (3) plans to engage in open activities for any future outputs pertaining to this project.

2. For each of the categories below, provide *representative examples* demonstrating how you have made research outputs resulting from this project openly accessible. If possible, provide the DOI and license terms under which the materials are available.

    - Open access articles
    - Open access books, book chapters, and/or monographs
    - Copies of your papers, chapters, monographs, or other published materials in institutional or disciplinary repositories
    - Preprints
    - Datasets
    - Software/Code
    - Materials/Reagents

- Preregistration plans
- Other outputs (please describe)

Additionally, it is important to include negative and null results, which could be covered in a variety of information formats.

UNIVERSITIES

Faculty Annual Report

1. For each of the categories below, provide *representative examples* demonstrating how (where appropriate) you have made outputs resulting from your research openly accessible. If possible, provide the DOI and license terms under which the materials are available.

    - Open access articles
    - Open access books, book chapters, and/or monographs
    - Copies of your papers, chapters, monographs, or other published materials in institutional or disciplinary repositories
    - Preprints
    - Datasets
    - Software/Code
    - Materials/Reagents
    - Preregistration plans
    - Other outputs (please describe)

Additionally, it is important to include negative and null results, which could be covered in a variety of information formats.

2. If known, describe how others have made use of these open research outputs, and include relevant DOIs if possible. This can include use in other disciplines and outside of academia.

3. Describe the impact that your openly available research outputs from this evaluation period have had from the research, public policy, pedagogic, and/or societal perspectives.

University Job Posting/Application

1. University XYZ values transparent, replicable, and reproducible research and open science principles (the open sharing of research outputs, including, but not limited to, open access and open data). How have you engaged in "open" activities during your career and how do you plan to do so in the future?

Or

2. University XYZ values transparent, replicable research and open science principles (the open sharing of research outputs, including, but not limited to, open access and open data). Describe the impact that your openly available research outputs have had from the research, public policy, pedagogic, and/or societal perspectives.

SENDING SIGNALS RUBRIC

This rubric complements the "Suggested Open Science Signaling Language" document produced by the same authors, which can be used by universities, agencies, philanthropies, and other stakeholders to highlight an organization's interest in open science activities at specific points of high leverage (such as grant applications, job postings). The rubric can be used by tenure and promotion committees, program managers, department chairs, hiring committees, and others tasked with evaluating the absolute and relative merits of responses to the signaling questions.

This workbook contains four sheets—one each with language pertaining specifically to articles, data, and other forms of research outputs at both application and reporting stages. The first sheet (Tables 1 and 2) is the amalgamated version, the second sheet (Tables 3 and 4) includes the articles version, and the third sheet (Tables 5 and 6) provides the data version. The fourth sheet (Tables 7 and 8) is the other output version that provides combined language encompassing all of these types of open science activities.

Please note that both the Sending Signals Language and the Sending Signals Rubric can be adapted to address the unique considerations, priorities, and norms of a specific community.

Table 1 Amalgamated Version – Application Stage

Application Stage (e.g., jobs, grants)	Beginning 1	Developing 2	Accomplished 3	Exemplary 4
Describe instances where you have engaged in "open" activities (such as making articles open access and sharing data/code according to FAIR principles), including representative examples	The researcher has not, in their recent research (<5 years), demonstrably engaged in open science practices such as making articles, data, and other research outputs openly available for access and reuse.	The researcher has sometimes engaged in open science practices. This is defined as occasionally making recent research (<5 years) available openly for access and reuse. Specific activities include (a) making at least one of their articles available in open access journals or repositories; (b) to the extent that the researcher has generated research data, making at least one of these datasets available in accessible repositories under adherence to the FAIR principles; and (c) to the extent that the researcher has generated research outputs beyond articles and data, making at least one of these materials openly available for access and reuse. Additionally, the researcher demonstrates at least some open science hygiene (e.g., use of DOIs, ORCID iDs, Creative Commons licenses).	The researcher has frequently engaged in open science practices. This is defined as often making recent research (<5 years) available openly for access and reuse. Specific activities include (a) making some (more than one, but less than most) of their articles available in open access journals or repositories; (b) to the extent that the researcher has generated research data, making some (more than one dataset, but less than most) of these data available in accessible repositories under adherence to the FAIR principles; and (c) to the extent that the researcher has generated research outputs beyond articles and data, making some (more than one, but less than most) of these materials openly available for access and reuse. Additionally, the researcher frequently demonstrates good open science hygiene (e.g., use of DOIs, ORCID iDs, Creative Commons licenses).	The researcher has consistently engaged in open science practices. This is defined as making the majority of recent research (<5 years) available openly for access and reuse. Specific activities include (a) making the majority of their articles available in open access journals or repositories; (b) to the extent that the researcher has generated research data, making the majority of these data available in accessible repositories under adherence to the FAIR principles; and (c) to the extent that the researcher has generated research outputs beyond articles and data, making the majority of these materials openly available for access and reuse. Additionally, the researcher consistently demonstrates good open science hygiene (e.g., use of DOIs, ORCID iDs, Creative Commons licenses).

continued

Table 1 Continued

Application Stage (e.g., jobs, grants)	Beginning 1	Developing 2	Accomplished 3	Exemplary 4
Provide examples of how your open research outputs have been used by others in your discipline, in other disciplines, and/ or outside of academia (include DOIs, if possible)	The researcher cannot provide qualitative and/or quantitative evidence that any of their recent (<5 years) open research outputs have been used by others.	The researcher can provide qualitative and/or quantitative evidence that at least one of their recent (<5 years) open research outputs has been used by others.	The researcher can provide qualitative and/or quantitative evidence that (a) some of their recent (<5 years) open research outputs have been used by others; and/or (b) a narrower range of their recent (<5 years) open research outputs have been used deeply within a specific community.	The researcher can provide qualitative and/or quantitative evidence that (a) a wide range of their recent (<5 years) open research outputs have been used by others; and/or (b) a narrower range of their recent (<5 years) open research outputs have been used deeply within a specific community.

Enumerate your plans to engage in open activities in the future	The researcher has not articulated a clear plan to make at least some research outputs (including, but not limited to, articles and data) available openly for access and reuse.	The researcher has articulated a clear plan to make at least some research outputs (including, but not limited to, articles and data) available openly for access and reuse. Specific activities include (a) making at least some of their articles available in open access journals or repositories; (b) to the extent that the researcher has generated research data, making most of these data available in accessible repositories under adherence to the FAIR principles; and (c) to the extent that the researcher has generated research outputs beyond articles and data, making at least some of these materials openly available for access and reuse. Additionally, the researcher has articulated a plan that demonstrates an awareness of at least some aspects of good open science hygiene (e.g., use of DOIs, ORCID iDs, Creative Commons licenses).	The researcher has articulated a clear plan to make most research outputs (including, but not limited to, articles and data) available openly for access and reuse. Specific activities include (a) making most of their articles available in open access journals or repositories; (b) to the extent that the researcher has generated research data, making most of these data available in accessible repositories under adherence to the FAIR principles; and (c) to the extent that the researcher has generated research outputs beyond articles and data, making most of these materials openly available for access and reuse. Additionally, the researcher has articulated a plan that demonstrates an intent to engage in good open science hygiene in most instances (e.g., use of DOIs, ORCID iDs, Creative Commons licenses).	The researcher has articulated a clear plan to make all appropriate research outputs (including, but not limited to, articles and data) available openly for access and reuse. Specific activities include (a) making their articles available in open access journals or repositories; (b) to the extent that the researcher has generated research data, making these data available in accessible repositories under adherence to the FAIR principles; and (c) to the extent that the researcher has generated research outputs beyond articles and data, making these materials openly available for access and reuse. Additionally, the researcher has articulated a clear and consistent plan to engage in good open science hygiene (e.g., use of DOIs, ORCID iDs, Creative Commons licenses).

NOTE: FAIR – Findability, Accessibility, Interoperability, and Reuse of digital assets; ORCID – Open Researcher and Contributor ID; DOI – Digital Object Identifier.

Table 2 Amalgamated Version – Reporting Stage

Reporting Stage (e.g., faculty tenure and promotion reviews, interim and final grant reports)	Beginning 1	Developing 2	Accomplished 3	Exemplary 4
For your work (related to this grant/during this time period), describe instances where you have engaged in "open" activities (such as making articles open access and sharing data/code according to FAIR principles), including representative examples	The researcher has not, in their research (for this project/period), demonstrably engaged in open science practices such as making articles, data, and other research outputs openly available for access and reuse.	The researcher has sometimes engaged in open science practices. This is defined as occasionally making research (for this project/period) available openly for access and reuse. Specific activities include (a) making at least one of their articles available in open access journals or repositories; (b) to the extent that the researcher has generated research data, making at least one of these datasets available in accessible repositories under adherence to the FAIR principles; and (c) to the extent that the researcher has generated research outputs beyond articles and data, making at least one of these materials openly available for access and reuse. Additionally, the researcher demonstrates at least some open science hygiene (e.g., use of DOIs, ORCID iDs, Creative Commons licenses).	The researcher has frequently engaged in open science practices. This is defined as often making research (for this project/period) available openly for access and reuse. Specific activities include (a) making some (more than one, but less than most) of their articles available in open access journals or repositories; (b) to the extent that the researcher has generated research data, making some (more than one dataset, but less than most) of these data available in accessible repositories under adherence to the FAIR principles; and (c) to the extent that the researcher has generated research outputs beyond articles and data, making some (more than one, but less than most) of these materials openly available for access and reuse. Additionally, the researcher frequently demonstrates good open science hygiene (e.g., use of DOIs, ORCID iDs, Creative Commons licenses).	The researcher has consistently engaged in open science practices. This is defined as making the majority of research (for this project/period) available openly for access and reuse. Specific activities include (a) making the majority of their articles available in open access journals or repositories; (b) to the extent that the researcher has generated research data, making the majority of these data available in accessible repositories under adherence to the FAIR principles; and (c) to the extent that the researcher has generated research outputs beyond articles and data, making the majority of these materials openly available for access and reuse. Additionally, the researcher consistently demonstrates good open science hygiene (e.g., use of DOIs, ORCID iDs, Creative Commons licenses).

For your work (related to this grant/during this time period), provide examples of how your open research outputs have been used by others in your discipline, in other disciplines, and/or outside of academia (include DOIs, if possible)	The researcher cannot provide qualitative and/or quantitative evidence that any of their open research outputs (for this project/period) have been used by others.	The researcher can provide qualitative and/or quantitative evidence that at least one of their open research outputs (for this project/period) has been used by others.	The researcher can provide qualitative and/or quantitative evidence that (a) some of their open research outputs (for this project/period) have been used by others; and/or (b) a narrower range of their open research outputs (for this project/period) have been used deeply within a specific community.	The researcher can provide qualitative and/or quantitative evidence that (a) a wide range of their open research outputs (for this project/period) have been used by others; and/or (b) a narrower range of their open research outputs (for this project/period) have been used deeply within a specific community.

continued

Table 2 Continued

Reporting Stage (e.g., faculty tenure and promotion reviews, interim and final grant reports)	Beginning 1	Developing 2	Accomplished 3	Exemplary 4
For your work (related to this grant/during this time period), enumerate your plans to engage in open activities in the future	The researcher has not articulated a clear plan to make at least some research outputs (including, but not limited to, articles and data) available openly for access and reuse.	The researcher has articulated a clear plan to make at least some research outputs (including, but not limited to, articles and data) available openly for access and reuse. Specific activities include (a) making at least some of their articles available in open access journals or repositories; (b) to the extent that the researcher has generated research data, making most of these data available in accessible repositories under adherence to the FAIR principles; and (c) to the extent that the researcher has generated research outputs beyond articles and data, making at least some of these materials openly available for access and reuse. Additionally, the researcher has articulated a plan that demonstrates an awareness of at least some aspects of good open science hygiene (e.g., use of DOIs, ORCID iDs, Creative Commons licenses).	The researcher has articulated a clear plan to make most research outputs (including, but not limited to, articles and data) available openly for access and reuse. Specific activities include (a) making most of their articles available in open access journals or repositories; (b) to the extent that the researcher has generated research data, making most of these data available in accessible repositories under adherence to the FAIR principles; and (c) to the extent that the researcher has generated research outputs beyond articles and data, making most of these materials openly available for access and reuse. Additionally, the researcher has articulated a plan that demonstrates an intent to engage in good open science hygiene in most instances (e.g., use of DOIs, ORCID iDs, Creative Commons licenses).	The researcher has articulated a clear plan to make all appropriate research outputs (including, but not limited to, articles and data) available openly for access and reuse. Specific activities include (a) making their articles available in open access journals or repositories; (b) to the extent that the researcher has generated research data, making these data available in accessible repositories under adherence to the FAIR principles; and (c) to the extent that the researcher has generated research outputs beyond articles and data, making these materials openly available for access and reuse. Additionally, the researcher has articulated a clear and consistent plan to engage in good open science hygiene (e.g., use of DOIs, ORCID iDs, Creative Commons licenses).

NOTE: FAIR – Findability, Accessibility, Interoperability, and Reuse of digital assets; ORCID – Open Researcher and Contributor ID; DOI – Digital Object Identifier.

57

Table 3 Articles Version – Application Stage

Application Stage (e.g., jobs, grants)	Beginning 1	Developing 2	Accomplished 3	Exemplary 4
Describe instances where you have engaged in making articles open access, including representative examples	The researcher has not, in their recent research (<5 years), demonstrably engaged in making articles openly available for access and reuse.	The researcher has sometimes engaged in open access practices. This is defined as occasionally making recent research articles (<5 years) available openly for access and reuse. Specific activities include (a) making at least one of their articles available in open access journals or repositories; and (b) demonstrating at least some open science hygiene (e.g., use of DOIs, ORCID iDs, Creative Commons licenses).	The researcher has frequently engaged in open access practices. This is defined as often making recent research articles (<5 years) available openly for access and reuse. Specific activities include (a) making some (more than one, but less than most) of their articles available in open access journals or repositories; and (b) frequently demonstrating good open science hygiene (e.g., use of DOIs, ORCID iDs, Creative Commons licenses).	The researcher has consistently engaged in open access practices. This is defined as making the majority of recent research articles (<5 years) available openly for access and reuse. Specific activities include (a) making the majority of their articles available in open access journals or repositories; and (b) consistently demonstrating good open science hygiene (e.g., use of DOIs, ORCID iDs, Creative Commons licenses).

continued

Table 3 Continued

Application Stage (e.g., jobs, grants)	Beginning 1	Developing 2	Accomplished 3	Exemplary 4
Provide examples of how your open access articles have been used by others in your discipline, in other disciplines, and/or outside of academia (include DOIs if possible)	The researcher cannot provide qualitative and/or quantitative evidence that any of their recent (<5 years) open access articles have been used by others.	The researcher can provide qualitative and/or quantitative evidence that at least one of their recent (<5 years) open access articles has been used by others.	The researcher can provide qualitative and/or quantitative evidence that (a) some of their recent (<5 years) open access articles have been used by others; and/or (b) a narrower range of their recent (<5 years) open access articles have been used deeply within a specific community.	The researcher can provide qualitative and/or quantitative evidence that (a) a wide range of their recent (<5 years) open access articles have been used by others; and/or (b) a narrower range of their recent (<5 years) open access articles have been used deeply within a specific community.
Enumerate your plans to engage in open access activities in the future	The researcher has not articulated a clear plan to make at least some research articles available openly for access and reuse.	The researcher has articulated a clear plan to make at least some research articles available openly for access and reuse. Specific activities include (a) making at least some of their articles available in open access journals or repositories; and (b) articulating a plan that demonstrates an awareness of at least some aspects of good open science hygiene (e.g., use of DOIs, ORCID iDs, Creative Commons licenses).	The researcher has articulated a clear plan to make most research articles available openly for access and reuse. Specific activities include (a) making most of their articles available in open access journals or repositories; and (b) articulating a plan that demonstrates an intent to engage in good open science hygiene in most instances (e.g., use of DOIs, ORCID iDs, Creative Commons licenses).	The researcher has articulated a clear plan to make all appropriate research articles available openly for access and reuse. Specific activities include (a) making their articles available in open access journals or repositories; and (b) articulating a clear and consistent plan to engage in good open science hygiene (e.g., use of DOIs, ORCID iDs, Creative Commons licenses).

NOTE: DOI – Digital Object Identifier; ORCID – Open Researcher and Contributor ID.

Table 4 Articles Version – Reporting Stage

Reporting Stage (e.g., faculty tenure and promotion reviews, interim and final grant reports)	Beginning 1	Developing 2	Accomplished 3	Exemplary 4
For your work (related to this grant/during this time period), describe instances where you have engaged in open access activities, including representative examples	The researcher has not, in their research (for this project/period), demonstrably engaged in making research articles openly available for access and reuse.	The researcher has sometimes engaged in open access practices. This is defined as occasionally making research articles (for this project/period) available openly for access and reuse. Specific activities include (a) making at least one of their articles available in open access journals or repositories; and (b) demonstrating at least some open science hygiene (e.g., use of DOIs, ORCID iDs, Creative Commons licenses).	The researcher has frequently engaged in open access practices. This is defined as often making research articles (for this project/period) available openly for access and reuse. Specific activities include (a) making some (more than one, but less than most) of their articles available in open access journals or repositories; and (b) frequently demonstrating good open science hygiene (e.g., use of DOIs, ORCID iDs, Creative Commons licenses).	The researcher has consistently engaged in open access practices. This is defined as making the majority of research articles (for this project/period) available openly for access and reuse. Specific activities include (a) making the majority of their articles available in open access journals or repositories; and (b) consistently demonstrating good open science hygiene (e.g., use of DOIs, ORCID iDs, Creative Commons licenses).

continued

Table 4 Continued

Reporting Stage (e.g., faculty tenure and promotion reviews, interim and final grant reports)	Beginning 1	Developing 2	Accomplished 3	Exemplary 4
For your work (related to this grant/during this time period), provide examples of how your open access articles have been used by others in your discipline, in other disciplines, and/or outside of academia (include DOIs, if possible)	The researcher cannot provide qualitative and/or quantitative evidence that any of their open access articles (for this project/period) have been used by others.	The researcher can provide qualitative and/or quantitative evidence that at least one of their open access articles (for this project/period) has been used by others.	The researcher can provide qualitative and/or quantitative evidence that (a) some of their open access articles (for this project/period) have been used by others; and/or (b) a narrower range of their open access articles (for this project/period) have been used deeply within a specific community.	The researcher can provide qualitative and/or quantitative evidence that (a) a wide range of their open access articles (for this project/period) have been used by others; and/or (b) a narrower range of their open access articles (for this project/period) have been used deeply within a specific community.

For your work (related to this grant/during this time period), enumerate your plans to engage in open access activities in the future	The researcher has not articulated a clear plan to make at least some research articles (including, but not limited to, articles and data) available openly for access and reuse.	The researcher has articulated a clear plan to make at least some research articles available openly for access and reuse. Specific activities include (a) making at least some of their articles available in open access journals or repositories; and (b) articulating a plan that demonstrates an awareness of at least some aspects of good open science hygiene (e.g., use of DOIs, ORCID iDs, Creative Commons licenses).	The researcher has articulated a clear plan to make most research articles available openly for access and reuse. Specific activities include (a) making most of their articles available in open access journals or repositories; and (b) articulating a plan that demonstrates an intent to engage in good open science hygiene in most instances (e.g., use of DOIs, ORCID iDs, Creative Commons licenses).	The researcher has articulated a clear plan to make all appropriate research articles available openly for access and reuse. Specific activities include (a) making their articles available in open access journals or repositories; and (b) articulating a clear and consistent plan to engage in good open science hygiene (e.g., use of DOIs, ORCID iDs, Creative Commons licenses).

NOTE: DOI – Digital Object Identifier; ORCID – Open Researcher and Contributor ID.

Table 5 Data Version – Application Stage

Application Stage (e.g., jobs, grants)	Beginning 1	Developing 2	Accomplished 3	Exemplary 4
Describe instances where you have engaged in open data activities (such as sharing data according to FAIR principles), including representative examples	The researcher has not, in their recent research (<5 years), demonstrably engaged in making data available for access and reuse according to FAIR principles.	The researcher has sometimes engaged in open data practices. This is defined as occasionally making research data (<5 years) available for access and reuse according to FAIR principles. Specific activities include (a) making at least one of their datasets available in accessible repositories under adherence to the FAIR principles; and (b) demonstrating at least some open science hygiene (e.g., use of DOIs, ORCID iDs, Creative Commons licenses).	The researcher has frequently engaged in open data practices. This is defined as often making recent research data (<5 years) available openly for access and reuse according to FAIR principles. Specific activities include (a) making some (more than one dataset, but less than most) of their research data available in accessible repositories under adherence to the FAIR principles; and (b) frequently demonstrating good open science hygiene (e.g., use of DOIs, ORCID iDs, Creative Commons licenses).	The researcher has consistently engaged in open data practices. This is defined as making the majority of recent research data (<5 years) available openly for access and reuse according to FAIR principles. Specific activities include (a) making the majority of their research data available in accessible repositories under adherence to the FAIR principles; and (b) consistently demonstrating good open science hygiene (e.g., use of DOIs, ORCID iDs, Creative Commons licenses).

Provide examples of how your open datasets have been used by others in your discipline, in other disciplines, and/or outside of academia (include DOIs, if possible)	The researcher cannot provide qualitative and/or quantitative evidence that any of their recent (<5 years) open datasets have been used by others.	The researcher can provide qualitative and/or quantitative evidence that at least one of their recent (<5 years) open datasets has been used by others.	The researcher can provide qualitative and/or quantitative evidence that (a) some of their recent (<5 years) open datasets have been used by others; and/or (b) a narrower range of their recent (<5 years) open datasets have been used deeply within a specific community.	The researcher can provide qualitative and/or quantitative evidence that (a) a wide range of their recent (<5 years) open datasets have been used by others; and/or (b) a narrower range of their recent (<5 years) open datasets have been used deeply within a specific community.
Enumerate your plans to engage in open data activities in the future	The researcher has not articulated a clear plan to make at least some research data available for access and reuse according to FAIR principles.	The researcher has articulated a clear plan to make at least some research data available for access and reuse according to FAIR principles. Specific activities include (a) making most of their research data available in accessible repositories under adherence to the FAIR principles; and (b) articulating a plan that demonstrates an awareness of at least some aspects of good open science hygiene (e.g., use of DOIs, ORCID iDs, Creative Commons licenses).	The researcher has articulated a clear plan to make most research data available for access and reuse according to FAIR principles. Specific activities include (a) making most of their research data available in accessible repositories under adherence to the FAIR principles; and (b) articulating a plan that demonstrates an intent to engage in good open science hygiene in most instances (e.g., use of DOIs, ORCID iDs, Creative Commons licenses).	The researcher has articulated a clear plan to make all appropriate research data available for access and reuse according to FAIR principles. Specific activities include (a) making their research data available in accessible repositories under adherence to the FAIR principles; and (b) articulating a clear and consistent plan to engage in good open science hygiene (e.g., use of DOIs, ORCID iDs, Creative Commons licenses).

NOTE: FAIR – Findability, Accessibility, Interoperability, and Reuse of digital assets; DOI – Digital Object Identifier; ORCID – Open Researcher and Contributor ID.

Table 6 Data Version – Reporting Stage

Reporting Stage (e.g., faculty tenure and promotion reviews, interim and final grant reports)	Beginning 1	Developing 2	Accomplished 3	Exemplary 4
For your work (related to this grant/during this time period), describe instances where you have engaged in open data activities (such as sharing data according to FAIR principles), including representative examples	The researcher has not, in their research (for this project/period), demonstrably engaged in making data available for access and reuse according to FAIR principles.	The researcher has sometimes engaged in open data practices. This is defined as occasionally making research data (for this project/period) available for access and reuse according to FAIR principles. Specific activities include (a) making at least one of their datasets available in accessible repositories under adherence to the FAIR principles; and (b) demonstrating at least some open science hygiene (e.g., use of DOIs, ORCID iDs, Creative Commons licenses).	The researcher has frequently engaged in open data practices. This is defined as often making research data (for this project/period) available openly for access and reuse according to FAIR principles. Specific activities include (a) making some (more than one dataset, but less than most) of their research data available in accessible repositories under adherence to the FAIR principles; and (b) frequently demonstrating good open science hygiene (e.g., use of DOIs, ORCID iDs, Creative Commons licenses).	The researcher has consistently engaged in open data practices. This is defined as making the majority of research data (for this project/period) available openly for access and reuse according to FAIR principles. Specific activities include (a) making the majority of their research data available in accessible repositories under adherence to the FAIR principles; and (b) consistently demonstrating good open science hygiene (e.g., use of DOIs, ORCID iDs, Creative Commons licenses).

Prompt				
For your work (related to this grant/during this time period), provide examples of how your open datasets have been used by others in your discipline, in other disciplines, and/or outside of academia (include DOIs if possible)	The researcher cannot provide qualitative and/or quantitative evidence that any of their open datasets (for this project/period) have been used by others.	The researcher can provide qualitative and/or quantitative evidence that at least one of their open datasets (for this project/period) has been used by others.	The researcher can provide qualitative and/or quantitative evidence that (a) some of their open datasets (for this project/period) have been used by others; and/or (b) a narrower range of their open datasets (for this project/period) have been used deeply within a specific community.	The researcher can provide qualitative and/or quantitative evidence that (a) a wide range of their open datasets (for this project/period) have been used by others; and/or (b) a narrower range of their open datasets (for this project/period) have been used deeply within a specific community.
For your work (related to this grant/during this time period), enumerate plans to engage in open activities in the future	The researcher has not articulated a clear plan to make at least some research data available for access and reuse according to FAIR principles.	The researcher has articulated a clear plan to make at least some research data available for access and reuse according to FAIR principles. Specific activities include (a) making most of their research data available in accessible repositories under adherence to the FAIR principles; and (b) articulating a plan that demonstrates an awareness of at least some aspects of good open science hygiene (e.g., use of DOIs, ORCID iDs, Creative Commons licenses).	The researcher has articulated a clear plan to make most research data available for access and reuse according to FAIR principles. Specific activities include (a) making most of their research data available in accessible repositories under adherence to the FAIR principles; and (b) articulating a plan that demonstrates an intent to engage in good open science hygiene in most instances (e.g., use of DOIs, ORCID iDs, Creative Commons licenses).	The researcher has articulated a clear plan to make all appropriate research data available for access and reuse according to FAIR principles. Specific activities include (a) making their research data available in accessible repositories under adherence to the FAIR principles; and (b) articulating a clear and consistent plan to engage in good open science hygiene (e.g., use of DOIs, ORCID iDs, Creative Commons licenses).

NOTE: FAIR – Findability, Accessibility, Interoperability, and Reuse of digital assets; DOI – Digital Object Identifier; ORCID – Open Researcher and Contributor ID.

Table 7 Other Outputs Version – Application Stage

Application Stage (e.g., jobs, grants)	Beginning 1	Developing 2	Accomplished 3	Exemplary 4
Describe instances where you have engaged in "open" activities beyond sharing articles and data, including representative examples	The researcher has not, in their recent research (<5 years), demonstrably engaged in making research outputs beyond articles and data openly available for access and reuse.	The researcher has (a) occasionally made recent (<5 years) research outputs beyond articles and data available openly for access and reuse; and (b) demonstrated at least some open science hygiene (e.g., use of DOIs, ORCID iDs, Creative Commons licenses).	The researcher has frequently made recent (<5 years) research outputs beyond articles and data available openly for access and reuse. Specific activities include (a) making some (more than one, but less than most) of these outputs available for access and reuse; and (b) frequently demonstrating good open science hygiene (e.g., use of DOIs, ORCID iDs, Creative Commons licenses).	The researcher has (a) consistently made the majority of recent (<5 years) research outputs beyond articles and data available openly for access and reuse; and (b) consistently demonstrated good open science hygiene (e.g., use of DOIs, ORCID iDs, Creative Commons licenses).

Provide examples of how your open research outputs beyond articles and data have been used by others in your discipline, in other disciplines, and/or outside of academia (include DOIs if possible)	The researcher cannot provide qualitative and/or quantitative evidence that any of their recent (<5 years) open research outputs beyond articles and data have been used by others.	The researcher can provide qualitative and/or quantitative evidence that at least one of their recent (<5 years) open research outputs beyond articles and data has been used by others.	The researcher can provide qualitative and/or quantitative evidence that (a) some of their recent (<5 years) open research outputs beyond articles and data have been used by others; and/or (b) a narrower range of their recent (<5 years) open research outputs beyond articles and data have been used deeply within a specific community.	The researcher can provide qualitative and/or quantitative evidence that (a) a wide range of their recent (<5 years) open research outputs beyond articles and data have been used by others; and/or (b) a narrower range of their recent (<5 years) open research outputs beyond articles and data have been used deeply within a specific community.
Enumerate your plans to engage in open activities in the future, beyond the open sharing of articles and data	The researcher has not articulated a clear plan to make at least some research outputs beyond articles and data available openly for access and reuse.	The researcher has articulated a clear plan to (a) make at least some research outputs beyond articles and data available openly for access and reuse; and (b) engage in at least some aspects of good open science hygiene (e.g., use of DOIs, ORCID iDs, Creative Commons licenses).	The researcher has articulated a clear plan to (a) make most research outputs beyond articles and data available openly for access and reuse; and (b) engage in good open science hygiene in most instances (e.g., use of DOIs, ORCID iDs, Creative Commons licenses).	The researcher has articulated a clear plan to (a) make all appropriate research outputs beyond articles and data available openly for access and reuse; and (b) engage in consistent good open science hygiene (e.g., use of DOIs, ORCID iDs, Creative Commons licenses).

NOTE: DOI – Digital Object Identifier; ORCID – Open Researcher and Contributor ID.

Table 8 Other Outputs Version – Reporting Stage

Reporting Stage (e.g., faculty tenure and promotion reviews, interim and final grant reports)	Beginning 1	Developing 2	Accomplished 3	Exemplary 4
For your work (related to this grant/during this time period), describe instances where you have engaged in "open" activities (beyond sharing articles and data), including representative examples	The researcher has not, in their research (for this project/period), demonstrably engaged in making research outputs beyond articles and data openly available for access and reuse.	The researcher has (a) occasionally made research outputs (for this project/period) beyond articles and data available openly for access and reuse; and (b) demonstrated at least some open science hygiene (e.g., use of DOIs, ORCID iDs, Creative Commons licenses).	The researcher has frequently made research outputs research (for this project/period) beyond articles and data available openly for access and reuse. Specific activities include (a) making some (more than one, but less than most) of these outputs available for access and reuse; and (b) frequently demonstrating good open science hygiene (e.g., use of DOIs, ORCID iDs, Creative Commons licenses).	The researcher has (a) consistently made the majority of research outputs research (for this project/period) beyond articles and data available openly for access and reuse; and (b) consistently demonstrated good open science hygiene (e.g., use of DOIs, ORCID iDs, Creative Commons licenses).

For your work (related to this grant/during this time period), provide examples of how your open research outputs beyond articles and data have been used by others, in your discipline, in other disciplines, and/or outside of academia (include DOIs, if possible)	The researcher cannot provide qualitative and/or quantitative evidence (for this project/period) that any of their open research outputs beyond articles and data have been used by others.	The researcher can provide qualitative and/or quantitative evidence that at least one of their open research outputs (for this project/period) beyond articles and data has been used by others.	The researcher can provide qualitative and/or quantitative evidence that (a) some of their open research outputs (for this project/period) beyond articles and data have been used by others; and/or (b) a narrower range of their open research outputs (for this project/period) beyond articles and data have been used deeply within a specific community.	The researcher can provide qualitative and/or quantitative evidence that (a) a wide range of their open research outputs (for this project/period) beyond articles and data have been used by others; and/or (b) a narrower range of their open research outputs (for this project/period) beyond articles and data have been used deeply within a specific community.
For your work (related to this grant/during this time period), enumerate your plans to engage in open activities beyond sharing articles and data in the future	The researcher has not articulated a clear plan to make at least some research outputs beyond articles and data available openly for access and reuse.	The researcher has articulated a clear plan to (a) make at least some research outputs beyond articles and data available openly for access and reuse; and (b) engage in at least some aspects of good open science hygiene (e.g., use of DOIs, ORCID iDs, Creative Commons licenses).	The researcher has articulated a clear plan to (a) make most research outputs beyond articles and data available openly for access and reuse; and (b) engage in good open science hygiene in most instances (e.g., use of DOIs, ORCID iDs, Creative Commons licenses).	The researcher has articulated a clear plan to (a) make all appropriate research outputs beyond articles and data available openly for access and reuse; and (b) engage in consistent good open science hygiene (e.g., use of DOIs, ORCID iDs, Creative Commons licenses).

Table 8 Notes:
- The rubric can and should be adapted to reflect the questions being asked of researchers (e.g., if a grant report form does not ask about data sharing, the data sharing elements of the rubric can be excised).
- The "Reporting" language can be customized for grant reporting vs. departmental reporting.
- Researchers who generate data with personal identifiable information (PII) or other sensitive details that cannot be openly shared may indicate as such in their response.
- While the FAIR (Findable, Accessible, Interoperable, Reusable) data principles support open research, data can be FAIR without being open. The FAIR principles can accommodate legitimate exceptions to open sharing practices such as data with PII, as mentioned above.
- "Other Outputs" include a range of research products such as the National Academies Roundtable on Aligning Incentives for Open Science list enumerated in VI. Reimagining Outputs Worksheet.
- DOI – Digital Object Identifier; ORCID – Open Researcher and Contributor ID.

III. GOOD PRACTICES PRIMERS[4]

Nicholas Gibson, John Templeton Foundation
Jerry Sheehan, National Institutes of Health
Stuart Buck, Formerly, Arnold Ventures
J. C. Burgelman, Vrije Universiteit Brussel
Anne-Marie Coriat, Wellcome
Anne Koralova, Helmsley Trust
Heather Pierce, Association of American Medical Colleges
Dawid Potgieter, Templeton World Charity Foundation
Greg Tananbaum, Open Research Funders Group

Many organizations, particularly those that perform or fund research, are in the information-gathering stage with respect to their open science policies and practices. These concise primers are intended to provide decision makers with a high-level overview of the *what's* and *how's* of open sharing of various research outputs. Each primer (1–2 pages) addresses a different output type, delving into exemplars, dependencies, resourcing, and a range of other considerations. The following drafts provide a sense of what the primers will encompass. They do not provide a detailed rationale for adopting an open science policy, an analysis of the barriers, or a comprehensive guide to implementation, including the pros and cons of various approaches.

ARTICLES

Relevance to Open Ecosystem

Unrestricted access to, and reuse of, published journal articles benefits the research community by facilitating the dissemination of new information, thus maximizing opportunities for that work to lead to new insights and discoveries.

Considerations

Among the key issues that organizations will wish to address in developing a policy to make articles open are the following:

---

[4] The views expressed are those of the authors and do not necessarily reflect the official policies or positions of their employing organizations.

- ***Fulfillment.*** Can researchers adhere to the policy by publishing in a fully open access journal, a "hybrid" journal (a subscription-based journal that allows authors to make individual articles open access immediately on payment of an article publication charge), or by posting a copy of a paper in an open, trusted repository? If the latter is permissible, must a certain version (e.g., version of record, approved manuscript) be posted?
- ***Timing.*** Does the policy require that the articles be made openly available immediately, or is some embargo (e.g., 6 months) permissible?
- ***Financial Support.*** Will the policy maker provide funding to defray costs of open access (e.g., article processing charges)? If so, is there a cap on the amount? Must the researcher explicitly account for these expenses at the time of project design? Is there a mechanism for the researcher to have such costs covered after grant close?
- ***Discoverability.*** How will potential readers discover the openly available content? Will it be picked up by major indexing services or be made available in leading disciplinary repositories?
- ***Licensing and Reuse.*** What type of licensing requirements will the policy include to facilitate reuse? Free to read, preferably permanent, is often the primary focus of open access policies, but reuse considerations (including, but not limited to, text and data mining) also merit consideration.

Approaches

The practical implementation of a policy requiring access to published articles can take different forms (see Box 1). Some policies require publication in an open access journal or a hybrid journal. This can introduce a modest restriction on researchers' choice of publication venue, although thousands of journals are open access or offer a hybrid option.

Some policies promote deposit of a copy of the paper (which may not be the final, formatted version, depending on publisher or funder requirements) in a trusted repository. As virtually all journals allow some form of self-archiving, this approach places fewer restrictions on authors (see Box 2). It does require authors to proactively identify and deposit the paper in an appropriate repository. Some journals will, however, deposit articles or final submitted manuscripts in a selected repository on behalf of authors.

SPARC (Scholarly Publishing and Academic Resources Coalition) maintains a succinct resource for tracking, comparing, and understanding

BOX 1
Examples of Open Access Policies Requiring Publication in Open Access Journals

The Bill & Melinda Gates Foundation and the Wellcome Trust require funded researchers to publish their articles in open access journals, with no embargo period.[a] The option to publish in hybrid journals is being phased out by both organizations in 2021.

---

[a] See https://www.gatesfoundation.org/how-we-work/general-information/open-access-policy and https://wellcome.org/news/wellcome-updates-open-access-policy-align-coalition-s.

BOX 2
Examples of Self-Archiving Open Access Policies

- All U.S. federal science funding agencies require submission of the author's final manuscript or final published article to a designated repository such as PubMed Central, with public access provided no later than 12 months after publication.[a]
- Harvard University is among the many universities that asks faculty to deposit a version of their articles ("the accepted author manuscript") in Harvard's institutional repository.[b]
- The Academic Senate of the University of California adopted a systemwide open access policy in 2013 designed to make research articles authored by faculty available to the public at no charge.[c]

---

[a] See https://obamawhitehouse.archives.gov/sites/default/files/microsites/ostp/ostp_public_access_memo_2013.pdf.
[b] See https://osc.hul.harvard.edu/policies.
[c] See https://osc.universityofcalifornia.edu/for-authors/open-access-policy.

U.S. federal funder article-sharing policies;[5] ROARMAP (Registry of Open Access Repository Mandates and Policies) provides similar information about funders and universities;[6] and the federal interagency group CENDI posts information about federal agency public-access policies.[7] These sites can be used to compare and contrast different approaches that stakeholders are taking to open access policies.

Resourcing

Once open policies are implemented, organizations can undertake a range of activities to manage them. At the low-touch end of the spectrum, organizations can require researchers to document how they intend to comply. Depending on internal resources, some organizations spot-check these plans, while others simply rely on the honor system. Other organizations take a more engaged approach, requiring proof of compliance from researchers and checking this against internal expectations and guidelines. Additionally, funders are increasingly able to rely on emerging research infrastructure such as author and funder registries to automate aspects of the reporting process. Organizations without open policies may view administration and compliance as daunting tasks. However, each organization can make its own appropriate determination about the resources it is able to devote to these activities. Compliance monitoring can often be embedded within other regular research-reporting processes without adding significant burden on researchers or administrative staff.

Next Steps

The Open Research Funders Group (ORFG) can provide support and insight into best practices and available resources.[8] The ORFG Incentivization Blueprint provides model language that can be adapted and adopted by funders and other organizations.[9] It offers a stepwise approach to deploying a policy that can grow to encompass not only open access articles but also data, code, and other research outputs.

---

[5] See http://researchsharing.sparcopen.org/articles.
[6] See https://roarmap.eprints.org.
[7] See https://www.cendi.gov/projects/Public_Access_Plans_US_Fed_Agencies.html.
[8] See http://www.orfg.org.
[9] See http://www.orfg.org/incentivization-blueprint.

DATA

Relevance to Open Ecosystem

The ability to independently confirm results and conclusions is critical for evaluating scientific rigor and informing future research activities. Openly shared data can support reanalysis and confirmation of research findings. They can also shed light on research that is not published, which can occur when tested hypotheses are not confirmed or research is considered unproductive, thereby mitigating publication bias and improving the efficiency of the research process, and can lead to novel lines of inquiry. In particular, shared data can be reused for new analyses, whether independently or in combination with other data.

Considerations

Several issues merit consideration by organizations developing open data policies, including the following:

- *Scope.* What data are needed for the independent verification of research results? Which data are most valuable to preserve for reuse? What is the appropriate balance between making available large volumes of raw data versus smaller amounts of more processed data?
- *Metadata.* What documentation and descriptive details are necessary to allow others to use the data properly and without confusion? How does the policy ensure that information about the methodology and procedures used to collect the data, details about codes, definitions of variables, variable field locations, frequencies, and the like are properly collected and disseminated? Are there disciplinary-specific metadata schemas that should be used to facilitate discovery and reuse?
- *Timing.* Starting with the baseline expectation that data underlying reported results will be made available concurrent with the posting of research findings, are there legitimate exceptions? Should researchers be given a period of exclusivity to analyze research data additional to those directly supporting reported findings before sharing them with the community? If data are

not reported in a publication, what is an appropriate time line for sharing the data?
- ***Financial Support.*** Who will provide funding to defray costs of preparing and/or depositing the data? What costs are recoverable? If so, is there a cap on the amount? Must the researcher explicitly account for these expenses at the time of project design?
- ***Licensing.*** What type of licensing requirements will the policy include to facilitate reuse of the data?
- ***Proprietary Software.*** To the extent that the data can only be accessed or analyzed through software that is not open source, what steps can be taken to reduce restrictions on its reuse?
- ***Data Management Plans.*** What support and guidance will the organization provide to help the researcher clearly articulate at the outset of a project what, how, and where data will be shared? What mechanisms are in place to ensure that the researcher adheres to the data management plan?
- ***Data Standards.*** For the study type in question, or for the field in which the work is centered, are there best practices for how the data should be formatted, to enable wider and more efficient reuse and interoperability?
- ***Preservation.*** What constitutes an appropriate deposit location for the data? Is there a repository that is appropriate for the subject matter in question, and/or has emerged within a specific research community as the default resource in that field? Is the repository secure, stable, and open for all to access?
- ***Discoverability.*** How will data be discoverable? Even if it is deposited in a particular repository, how will other possible users know where to look? Will the data be assigned a unique persistent identifier, and will that identifier be promulgated through related publications?
- ***Privacy/Confidentiality.*** Some datasets may contain human subject details that cannot be fully disseminated, due to the Health Insurance Portability and Accountability Act of 1996 (HIPAA) (Public Law 104-191; 104th Congress), the Family Educational Rights and Privacy Act (FERPA) (20 U.S.C. § 1232g; 34 C.F.R. Part 99), the European Union's General Data Protection Regulation 2016/679 (EU GDPR) (O.J. L. 119, 04.05.2016; cor. O.J. L. 127, 23.5.2018), or other privacy restrictions. Such datasets, however, can often be shared after anonymization or deidentifica-

tion techniques (including adding statistical noise, suppression of small cells, etc.), or under protected mechanisms such as a virtual data warehouse accessible only with a confidentiality agreement in place. How will such datasets be handled in a way that maximizes sharing while protecting privacy? Can analytic opportunities be made openly available while the confidential aspects of the data remain restricted?
- *Compliance monitoring.* How can compliance with data management and sharing requirements/expectations be easily monitored, for example, by funders, other institutions, or individuals?

Approaches

One common approach to facilitate data sharing is to develop policies requiring data to be findable, accessible, interoperable, and reusable, that is, to meet the Findability, Accessibility, Interoperability, and Reuse of digital assets (FAIR) data principles. While data can be FAIR without necessarily being publicly open, the FAIR principles broadly support open science. Specific definitions and operationalizations of each of these principles, together with practical guidance on how to satisfy each requirement, have been prepared by the GO FAIR Initiative.[10] To render data FAIR, metadata and datasets should be prepared in a standardized, descriptive manner that makes it easier for both humans and machines to find and use.

With respect to data accessibility, a common rule of thumb in the open science community is that data should be shared in a manner that promotes reuse and transparency while recognizing that certain safeguards may be required to protect sensitive information that could compromise subject privacy or other norms and regulations. While the default position needs to shift to "open," legitimate restrictions on access need to be taken into account.

Many U.S. federal science agencies require researchers to submit a data management plan either as part of a grant application or before issuing an award. These plans provide general information about the types of data to be collected in a research study, the repository into which they will be deposited, and the time lines and other conditions of access. For certain types of research studies, federal science agencies have developed more

---

[10] See https://www.go-fair.org/fair-principles.

specific guidance or requirements (see the National Institutes of Health [NIH] example in Box 3).

Some organizations, such as the National Science Foundation, provide a general set of guidelines on data sharing, articulating to researchers that they are expected to share their data with their peers under reasonable cir-

---

BOX 3
Examples of Open Data Policies

- The National Institutes of Health (NIH) issued a Data Management and Sharing Policy that applies to all data generated by funded research[a] as well as specific policies that apply to genomic data, clinical trial data, and other specific research programs and data types.[b] NIH has also provided information for selecting a data repository.[c]
- The American Heart Association requires grant applicants to include a data-sharing plan as part of the application process. Any research data that are needed for independent verification of research results must be made freely and publicly available within 12 months of the end of the funding period (and any no-cost extension).
- The European Open Science Cloud (EOSC) has developed a strategic implementation plan for the creation of a data commons housing interoperable, machine-readable data across domains, consistent with FAIR (findable, accessible, interoperable, and reusable) principles.[d]
- The Yale University Open Data Access (YODA) Project facilitates clinical trial data access to promote independent analyses of the data. It also provides a formal vetting of the data to ensure consistency with informed consent and confidentiality requirements.[e]

---

[a] See https://grants.nih.gov/grants/guide/notice-files/NOT-OD-21-013.html.

[b] For genomic data, see https://osp.od.nih.gov/scientific-sharing/genomic-data-sharing; for clinical trial data, see https://grants.nih.gov/policy/clinical-trials/reporting/understanding/nih-policy.htm; and for other specific research programs and data types, see https://www.nlm.nih.gov/NIHbmic/nih_data_sharing_policies.html.

[c] See https://grants.nih.gov/grants/guide/notice-files/NOT-OD-21-016.html.

[d] See https://ec.europa.eu/info/research-and-innovation/strategy/goals-research-and-innovation-policy/open-science/european-open-science-cloud-eosc_en.

[e] See https://yoda.yale.edu.

cumstances.[11] Others, such as the NIH, have overarching data management and sharing policies that apply to all funded research, while also having more focused policies that provide explicit guidance as to the timing, licensing, and dissemination of data of particular types (e.g., genomic data) or associated with particular research programs (e.g., the Cancer Moonshot).[12]

Resourcing

For data specifically, it is important to ensure that appropriate metadata and documentation are provided so that datasets are properly contextualized. Organizations will also benefit from in-house or outsourced expertise to assess the appropriateness of data management plans and informed consents, to ensure these allow data sharing to the extent that the organization desires.

Once open policies are implemented, organizations can undertake a range of activities to manage them. At the low-touch end of the spectrum, organizations can require researchers to document how they intend to comply. Depending on internal resources, some organizations spot-check these plans, while others simply rely on the honor system. Other organizations take a more engaged approach, requiring proof of compliance from researchers and checking this against internal expectations and guidelines. Additionally, funders are increasingly able to rely on emerging research infrastructure, such as author and funder registries, to automate aspects of the reporting process. Organizations without open policies may view administration and compliance as daunting tasks. However, each organization can make its own appropriate determination about the resources it is able to devote to these activities.

Next Steps

There are a range of resources that can contribute to a detailed understanding of policy options and approaches, including the following:

- GO FAIR provides a starter kit with a wealth of information on data management plans, license options, and repositories.[13]

---

[11] See https://www.nsf.gov/pubs/policydocs/pappg19_1/pappg_11.jsp#XID4.
[12] See https://www.cancer.gov/research/key-initiatives/moonshot-cancer-initiative/funding/public-access-policy#requirement.
[13] See https://www.go-fair.org/resources/rdm-starter-kit.

- The Transparency and Openness Promotion (TOP) Guidelines provide sample language for three levels of open data policies.[14] This wording can be adapted and adopted to suit the specific circumstances of various organizations.
- The Open Research Funders Group Incentivization Blueprint offers sample open data policy language that can be adapted for a range of use cases.[15]
- The American Heart Association's website contains a detailed FAQ page that articulates questions commonly asked by researchers subject to an open data policy.[16]
- The DMPTool site is an excellent resource for both browsing the data policies of hundreds of organizations and generating data management plans to fit a range of requirements and circumstances.[17]
- NIH is developing various resources to assist researchers in complying with its Data Management and Sharing Policy, including clarifications about the contents of a data management and sharing plan, selection of data repositories, and allowable costs.[18]

---

[14] See https://osf.io/bcj53.
[15] See http://www.orfg.org/incentivization-blueprint.
[16] See https://professional.heart.org/en/research-programs/aha-research-policies-and-awardee-hub/open-science-frequently-asked-questions#:~:text=The%20AHA%20open%20data%20policy%20requires%20any%20data%20needed%20for,but%20the%20most%20exceptional%20circumstances.
[17] See https://dmptool.org.
[18] See https://grants.nih.gov/grants/guide/notice-files/NOT-OD-21-013.html.

APPENDIX C

PROTOCOLS AND PREREGISTRATION ANALYSIS PLANS

Relevance to Open Ecosystem

Unreported flexibility in data analysis can reduce the credibility of reported results and invalidate common tools of statistical inference. By submitting a detailed study protocol and statistical analysis plan to a public registry prior to conducting the work (i.e., preregistering with an analysis plan), the scientist makes a clearer distinction between planned hypothesis tests (i.e., confirmatory tests) and unplanned discovery research (i.e., screening or exploratory research). Preregistration of laboratory protocols—detailed descriptions of the methods used in the experiment, including equipment and reagents—is becoming more common and facilitates replicability. Preregistration is particularly important for studies that make an inferential claim from a sampled group or population, as well as studies that are reporting and testing hypotheses. After a project is completed, protocols and preregistration analysis plans can be used in conjunction with the final study and analysis by researchers seeking to replicate, reproduce, and build upon findings.

Considerations

- ***Scope.*** Should preregistration address the study protocol (how a study or experiment will be conducted), the laboratory protocol (detailed description of methods), the analysis plan (how the collected data will be organized and evaluated), or all three? Of primary interest in ensuring the integrity of the research outcome is information about the prespecified outcome measures/endpoints. However, decisions made during analysis can also affect the integrity of the reported findings, so many registries encourage preregistration of both.
- ***Documentation.*** Should preregistration include disclosure of the full-study protocol or just summary information about the protocol and statistical analysis plan? Submission of summary information can be more time consuming, but it also allows for structured data entry to facilitate searching and cross-study comparison. If a summary is submitted, then what specific information needs to be provided?
- ***Data Privacy.*** Protocols and analysis plans can contain proprietary or other protected information (e.g., names of study person-

nel). To what extent can information be redacted without undermining the benefits of access? The desire to promote meaningful preregistration must be balanced against the provision of necessary protections/redactions of information.

- **Deposit Location.** Where and how should a scientist register their protocol and/or analysis plan? There are a limited number of established public repositories. For clinical trials of health-related interventions, NIH's ClinicalTrials.gov is the default system.[19] Within the social, behavioral, and preclinical sciences, the Open Science Framework is becoming a default registry.[20] Some public repositories tend to be disciplinarily focused.
- **Timing.** How long before or after a study begins must it be registered? When should a preregistration be updated? Earlier may be better, but additional information may be needed about its status (e.g., has Institutional Review Board approval been received). The timing of an update is also linked to the degree to which a change has implications on the full preregistration (e.g., challenges in recruiting a full sample may necessitate moving from a single cohort to a multicohort design). Protocols shared at study initiation can more clearly establish a project's aims and plan. Does the registry support time-stamped versioning?
- **Discoverability.** Are preregistrations automatically made public after a fixed period of time? Does the registry support public searches for preregistrations?
- **Scope.** To date, the majority of registries are for causal impact studies, typically carried out either in a small-scale experiment or a large randomized clinical/field trial. However, there may be a strong rationale to consider preregistering exploratory studies at the time of funding or at the beginning of a study so as to capture strong theory-driven exploratory questions as opposed to post hoc "fishing" analyses.
- **Results.** To what extent should a funder require the ultimate posting of a study's results in a way that can be compared to whatever was preregistered? Federal law requires the posting of results at ClinicalTrials.gov for certain clinical trials; should this be a broader expectation?

---

[19] See https://clinicaltrials.gov.
[20] See https://osf.io.

Approaches

There are a range of different preregistration locations available, primarily driven by discipline. All NIH-funded clinical trials and most clinical trials of Food and Drug Administration (FDA) regulated drugs, biologics, and devices must be preregistered at NIH's ClinicalTrials.gov not later than 21 days after first recruitment. Summary information is provided in a highly structured format. Final protocols for NIH-funded clinical trials and most FDA-regulated clinical trials of drugs, biologics, and devices must be submitted to NIH's ClinicalTrials.gov as part of summary data reporting after a trial has completed. These policies also require that the statistical analysis plan be submitted if it is not considered part of the protocol. (See Box 4 for examples of preregistration and protocols policies.)

BOX 4
Examples of Preregistration and Protocols Policies

- The Chan Zuckerberg Initiative (CZI) requires grantees to make laboratory protocols publicly available and has nurtured dedicated protocol communities of CZI-funded investigators.[a]
- The American Economic Association (AEA) encourages researchers to register their randomized controlled trials (including research designs and analysis plans) in the AEA Randomized Controlled Trials (RCT) Registry.[b]
- CHDI Foundation has established an Independent Statistical Standing Committee (ISSC) to provide unbiased evaluation and expert advice on developing protocols and statistical analysis plans, and evaluation of prepared study protocols.[c]
- Arnold Ventures requires all funded empirical studies that involve statistical inference to be preregistered before the start of intervention or data collection on the Open Science Framework.[d]

---

[a] See https://cziscience.medium.com/power-to-the-protocols-388fe92001be and https://www.protocols.io/workspaces/neurodegeneration-method-development-community1.
[b] See https://www.socialscienceregistry.org.
[c] See https://chdifoundation.org/independent-statistical-standing-committee.
[d] See https://www.arnoldventures.org/guidelines-for-investments-in-research and https://osf.io.

Other disciplines have their own community-promoted repositories. Researchers carrying out causal studies in education have the opportunity to preregister their work in the Registry of Efficacy and Effectiveness Studies.[21] Researchers in the social, behavioral, and cognitive sciences often use the Open Science Framework platform.[22] The Registry for International Development Impact Evaluations hosts impact evaluations related to development in low- and middle-income countries.[23]

Resourcing

Organizations considering preregistration will need to consider whether resources are needed to support a preregistration repository for collecting preregistration reports and protocols. It is also important that there is a transparent link among any disseminated findings (preprints, articles, etc.), data, and preregistrations to determine whether there are significant deviations from the intended analysis.

Organizations and publishers will also need to ascertain how to indicate where preregistration records and protocol information exist for a published article. Multiple publishers and other organizations offer modalities for publishing study protocols, laboratory protocols, and registered reports. To be most effective, preregistrations and protocols should be closely linked to associated publications and other study information so that they can be easily discovered and accessed by those examining the study results.

Next Steps

The TOP Guidelines provide sample language for three levels of policies for study preregistration and analysis plan preregistration.[24] This wording can be adapted and adopted to suit the specific circumstances of a range of organizations. The TOP recommendations include (1) disclosing whether work was preregistered or not, (2) verifying that any preregistered work adheres to the prespecified plans, and (3) requiring preregistration for relevant research studies (typically inferential and hypothesis-testing work).

---

[21] See https://sreereg.icpsr.umich.edu/sreereg.
[22] See https://osf.io/prereg.
[23] See https://ridie.3ieimpact.org.
[24] See https://osf.io/bcj53.

The Center for Open Science provides multiple resources on how to preregister studies and analytic plans, including templates.[25] NIH provides a number of resources to facilitate the development of protocols, including the National Institutes of Health e-Protocol Writing Tool and protocol templates for clinical trials and behavioral/social science research.[26]

---

[25] See https://www.cos.io/initiatives/prereg and https://osf.io/zab38/wiki/home/?view.
[26] See https://e-protocol.od.nih.gov/#/home and https://grants.nih.gov/policy/clinical-trials/protocol-template.htm.

REGISTERED REPORTS

Relevance to Open Ecosystem

Peer review of study protocols with analysis plans, along with dissemination of findings regardless of outcome, addresses publication bias against null results. It also provides the benefits of preregistration by making a clearer distinction between hypothesis tests and discovery research. By submitting funded studies to journals as a registered report, the scientist improves study planning, increases study rigor, and improves scientific credibility. Funders who support this process anticipate that peer-review feedback could change study processes that result in budget changes and are prepared to consider such amendments in response to journal reviewer feedback. Funders can also partner with journals to coordinate review for funding and publishing decisions.

Considerations

- ***Scope.*** Registered reports are most appropriate for specific experiments or studies, not for grants that fund a research program over several years. Such grants could still include one or more registered reports, but it would likely not cover the entire program.
- ***Research Scope.*** Registered reports are best for studies that test hypotheses and in disciplines that could suffer from publication bias (typically against null results). Registered reports are not appropriate for purely exploratory or discovery science, until those studies are ready to use traditional hypothesis tests.
- ***Timing.*** By design, registered reports include additional time at the beginning of a project. Project plans should account for this. Additional time devoted to peer review in the early stages of the project is also required to ensure that the study methods are as rigorous as possible and that results will be disseminated regardless of outcome.

Approaches

There are a number of ways in which an organization can promote registered reports. On the low end of engagement, a funder or agency can ask grantees to specifically state whether all or part of the work would be

> **BOX 5**
> **Examples of Funders Encouraging/ Requiring Registered Reports**
>
> - The Flu Lab is partnering with PLOS and the Center for Open Science to promote replications and registered reports of influenza research.
> - Cancer Research UK is collaborating with the journal *Nicotine & Tobacco Research* on an integrated review process for grant proposals and preregistered reports.[a]
>
> [a] See https://academic.oup.com/ntr/article/19/7/773/3106460.

appropriate for a registered report. This will remind grantees that registered reports are a valued addition to a proposed study. Principal investigators can be encouraged to notify their communities—via social media, their websites, CVs, and other appropriate channels—when their precollection hypotheses and data analysis plans have been reviewed and registered. Organizations may also wish to educate researchers on the benefits of registered reports, particularly researchers in domains where the practice is not currently widespread.

For specific grants, programs, or initiatives where projects are appropriate for the format, agencies and funders may elect to make registered report submissions to a journal before data collection a requirement. If a study does not receive an in-principle acceptance offer from a journal, the plan can still be preregistered by the authors on a platform like the Open Science Framework and submitted for publication after the study is completed.

Some funders are partnering directly with discipline-appropriate journals to integrate the registered reports model in the grant application process. One example is the Children's Tumor Foundation,[27] which is partnering with the journal *PLOS ONE* to concurrently evaluate grant proposals and the ethics and rigor of the experimental design. Accepted proposals will simultaneously receive both funding and a commitment to publication of the study results in *PLOS ONE*. (See Box 5.)

---

[27] See https://grants.nih.gov/policy/clinical-trials/protocol-template.htm.

Resourcing

Given the relative novelty of registered reports, organizations may need to educate grantees about the merits and mechanics of this approach. Organizations that seek to integrate grant proposals and registered reports will need to establish a review process that allows for independent evaluation of the latter along a timescale and workflow that supports the former. This may also require negotiation of a direct partnership with a journal or publisher.

Absent this type of embedded relationship, researchers may require guidance to evaluate the growing number of journals that accept and publish registered reports. The Comparison of Registered Reports site provides an interactive tool to assist in this process.[28] Policies that require registered reports will also require some form of monitoring, ranging from spot-checking to soliciting proof of compliance.

Next Steps

The Center for Open Science provides a comprehensive registered reports resource,[29] including FAQs, workflow suggestions, and other foundational materials. The Center for Open Science also provides a simple Q&A tutorial to assist authors in the drafting of registered reports.[30] The Open Science Framework provides a searchable database of registered reports across a range of disciplines.[31] These may offer useful guidance to better understand the core elements of a well-constructed registered report.

---

[28] See https://katiedrax.shinyapps.io/cos_registered_reports.
[29] See https://www.cos.io/initiatives/registered-reports.
[30] See https://osf.io/93znh/?_ga=2.100491997.298846709.1580837996-1159488863.1580234077.
[31] See https://osf.io/registries/discover?provider=OSF&type=Registered%20Report%20Protocol%20Preregistration.

SOFTWARE AND CODE

Relevance to Open Ecosystem

Research projects may generate code that is used as a means to run, analyze, or interpret research data. The ability to independently confirm results and conclusions is critical for evaluating scientific rigor and informing future research activities. To extract maximum value from research findings and available data, any code deployed to process these data must therefore be widely and freely available. Research findings are not fully open unless the tools necessary to understand and test them are also made available. Research projects may also generate software that is the product of the project rather than the byproduct, a specified deliverable designed to perform a specific task. Making the underlying code for this type of research output open source can encourage collaboration, further development, community engagement, and enhanced return on funders' investment.

Considerations

As organizations develop open science policies pertaining to code and software, among the issues they must consider are the following:

- *Software/Code Maintenance.* What are the expectations for the duration and extent to which code should be kept up to date? Should the version used to produce the reported findings be maintained?
- *Proprietary Software.* To the extent that some or all of the code base upon which an experiment relies is not open source, what steps can be taken to reduce restrictions on its reuse?
- *Timing.* Does the policy require that the code or software be made openly available immediately upon the posting of research findings (e.g., publication of an article, deposit of a dataset), or is some embargo (e.g., 6 months) permissible? If research findings are not published or posted, should code and software be made publicly available no later than grant close?
- *Financial Support.* Will the policy maker provide funding to defray costs of preparing and/or depositing the code or software? If so, is there a cap on the amount? Must the researcher explicitly account for these expenses at the time of project design? If code

or software is made publicly available after the conclusion of the grant, does the grantee have a mechanism to request additional financial support?
- **Licensing.** What type of licensing requirements will the policy include to facilitate reuse? Do the grantee and/or the funder retain any stake in the intellectual property?
- **Metadata.** What documentation and descriptive details are needed to understand and execute the code or run the software program? How will the computational environment in which software or code was originally executed be described and archived? Should researchers establish virtual environments (e.g., Docker)?
- **Preservation.** What constitutes an appropriate deposit location for the code or software? Is there a repository that is appropriate for the subject matter in question and/or has emerged within a specific research community as the default resource in that field? Is the repository secure, stable, and open for all to access? Does the repository assign persistent digital identifiers to code?

Approaches

The TOP Guidelines advise that researchers should "provide program code, scripts for statistical packages, and other documentation sufficient to allow an informed researcher to precisely reproduce all published results ... through a trusted digital repository."[32] More funder-specific TOP guidance may be found at https://www.cos.io/initiatives/top-funders.

Some agencies within the U.S government use open source code as a matter of policy. For example, the Consumer Financial Protection Bureau unequivocally states, "When we build our own software or contract with a third party to build it for us, we will share the code with the public at no charge."[33] Other agencies, such as the Department of Education, make the source code for their prominent public-facing initiatives (in ED's case, the College Scorecard)[34] openly available. Both of these organizations deposit these research outputs (software as a product, not a byproduct, of the grant) on GitHub. When code is developed to interpret or analyze research findings (code as a secondary output of the grant), organizations such as

---

[32] See https://osf.io/bcj53.
[33] See https://www.consumerfinance.gov/about-us/blog/the-cfpbs-source-code-policy-open-and-shared.
[34] See https://collegescorecard.ed.gov.

> **BOX 6**
> **Examples of Open-Code and Software Policies**
>
> - NASA's Earth Science Data Systems (ESDS) Program requires that all software developed through research and technology awards be made available to the public as open source.[a] All funding proposals must include software development plans that are vetted as part of the application process.
> - The U.S. government's Federal Source Code Policy includes a pilot program that "requires agencies, when commissioning new custom software, to release at least 20 percent of new custom-developed code as Open Source Software for three years."[b]
> - Several learned societies that publish flagship disciplinary journals, including the American Geophysical Union and the American Astronomical Society, require or strongly encourage authors to make openly available any code used to generate results or analyses reported in their papers.[c]
>
> [a] See https://earthdata.nasa.gov/collaborate/open-data-services-and-software/esds-open-source-policy.
> [b] See https://www.cio.gov/2016/08/11/peoples-code.html.
> [c] See https://www.agu.org/Publish-with-AGU/Publish/Author-Resources/Policies/Data-policy and https://journals.aas.org/news/policy-statement-on-software.

the Wellcome Trust typically require the code to be shared at the time the primary research is published.[35] (See Box 6 for examples of open-code and software policies.)

Resourcing

For code specifically, some technical expertise may be required to ensure that the code and software are operable and can be accessed and used by the wider community.

Once open policies are implemented, organizations can undertake a range of activities to manage them. At the low-touch end of the spectrum, organizations can require researchers to document how they intend to

---
[35] See https://wellcome.org/news/our-new-policy-sharing-research-data-what-it-means-you.

comply. Depending on internal resources, some organizations spot-check these plans, while others simply rely on the honor system. Other organizations take a more engaged approach, requiring proof of compliance from researchers and checking this against internal expectations and guidelines.

Next Steps

The TOP Guidelines provide sample language for three levels of open-code policies.[36] This wording can be adapted and adopted to suit the specific circumstances of a range of organizations. For a deeper dive into policy formulation, interested parties can download the National Academies of Sciences, Engineering, and Medicine's report *Open Source Software Policy Options for NASA Earth and Space Sciences*.[37] This comprehensive document provides a deep dive into the established approaches, best practices, and practical considerations that can help effectively shape an open code policy.

---

[36] See https://osf.io/bcj53.
[37] See https://www.nap.edu/catalog/25217.

APPENDIX C   *93*

IV. OPEN SCIENCE BY THE NUMBERS INFOGRAPHIC

Open Science by the Numbers

Open Science posits that research has its widest impact and is most trustworthy when all of its elements (including articles, data, protocols, and code) can be openly accessed, tested, and built upon.

Researchers estimate that $3.2 trillion in economic output could be added to global GDP through Open Data across all sectors, with scientific and scholarly data playing an important role.[1]

The Symbiota open source platform, funded by NSF, hosts **37 million biological records** from 766 universities, museums, and research organizations.[2]

700 Global Open Data for Agriculture and Nutrition (GODAN) is an open data sharing initiative drawing on the participation of **over 700 private and public sector, nonprofit, and academic organizations** with the goal of developing solutions to global hunger.[3]

1 month from first reported COVID-19 case to genetic sequencing, rapidly expedited by open science and data sharing

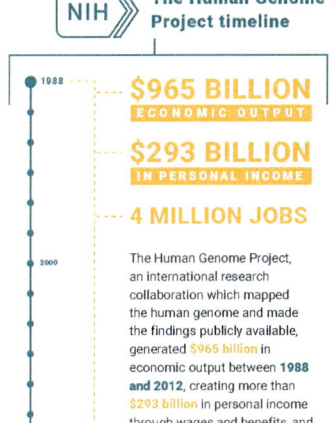

The Human Genome Project, an international research collaboration which mapped the human genome and made the findings publicly available, generated $965 billion in economic output between **1988** and **2012**, creating more than $293 billion in personal income through wages and benefits, and nearly 4 million jobs.[4]

In 2019, **31% of all journal articles were available as Open Access**, and 52% were viewed Open Access articles.[5]

Authors: Derrick Anderson, Arizona State University; Rachel Bruce, UK Research and Innovation; Ashley Farley, Bill & Melinda Gates Foundation; Robert Hanisch, National Institute of Standards and Technology; Greg Tananbaum, Open Research Funders Group; Thomas Wang, American Heart Association/University of Texas Southwestern Medical Center.
The views expressed are those of the authors and do not necessarily reflect the official policies or positions of their employing organizations.

Sources:
1. https://www.omidyar.com/sites/default/files/file_archive/insights/ON%20(Report_061114_FNL.pdf
2. http://symbiota.org/docs/
3. https://www.usda.gov/media/blog/2019/07/13/open-data-enabling-fact-based-data-driven-decisions
4. https://sparcopen.org/impact-story/human-genome-project/
5. https://www.biorxiv.org/content/10.1101/795310v1

V. OPEN SCIENCE SUCCESS STORIES DATABASE[38]

Derrick Anderson, Arizona State University
Greg Tananbaum, Open Research Funders Group

The Open Science Success Stories Database compiles articles, perspectives, case studies, news stories, and other materials that demonstrate the myriad ways in which open science benefits researchers and society alike.[39]

Scientists, scholars, librarians, department chairs, university administrators, philanthropic program officers, government agency representatives, policy makers, publishers, journalists, and other stakeholders can use the curated resources to understand how open science is positively influencing specific disciplines and communities, as well as how these lessons can be applied to the global scientific endeavor.

The database is being developed by Arizona State University in collaboration with the Open Research Funders Group. An initial version is being made available as part of the background material for the November 5, 2020, National Academies workshop on Developing a Toolkit for Fostering Open Science Practices.

---

[38] The views expressed are those of the authors and do not necessarily reflect the official policies or positions of their employing organizations.
[39] See https://projectopen.io.

VI. REIMAGINING OUTPUTS WORKSHEET[40]

Boyana Konforti, Formerly, Howard Hughes Medical Institute
Elizabeth Albro, U.S. Department of Education
Anurupa Dev, Association of American Medical Colleges
Josh Greenberg, Alfred P. Sloan Foundation
Ross Mounce, Arcadia Fund
Brian Quinn, Robert Wood Johnson Foundation
Greg Tananbaum, Open Research Funders Group
Richard Wilder, Coalition for Epidemic Preparedness Innovations

The following table (organized alphabetically) represents the authors' perspective about the range of research products that should be accounted for as the science community thinks about the behaviors and activities that should be rewarded. What are the outputs that are consistent with the values the science community collectively espouses? What outputs encourage open dialog and the tackling of big questions, build upon and enhance the work of others, and advance the research endeavor? As the community enumerates these research products, what considerations must be contemplated and addressed to create appropriate alignment between values and activities? The authors believe it will be crucial to ensure that the science community takes an expansive view of the types of research products that should be "open"—available for access and reuse without gatekeeping or payment.

---

[40] The views expressed are those of the authors and do not necessarily reflect the official policies or positions of their employing organizations.

Reimagining Outputs Worksheet Table

Research Output Type	Exemplar Open Practices	Importance to Open Ecosystem	Concerns/Considerations
Articles	All primary research articles should be made immediately available (open access with no embargo period) and reusable via an expansive license such as CC BY.	Unrestricted access to, and reuse of, published articles benefit the research community by facilitating the discovery of new information, thus maximizing opportunities for that work to lead to new insights and discoveries.	• Free to read is often the primary focus of open access policies, but reuse considerations (including, but not limited to, text and data mining) also merit consideration. • Distinctions between versions (version-of-record accepted manuscript) may be more important within certain disciplines.
Code and Software	To the greatest extent allowable by copyright, all software, code, lab notebooks, and executables necessary to independently verify research results should be curated and made freely available in an open repository no later than the publication of the first paper running this code.	The independent confirmation of results and conclusions is critical for understanding scientific soundness and informing future research activities. To extract maximum value from research findings, both the raw data that underpin the results and any code deployed to process these data must be widely and freely available to any interested party. Succinctly, research findings are not fully open unless the tools necessary to understand and test them are also made available.	Stewardship/ownership of repositories—ensuring these are open and sustainable.

Commentaries and Analyses	Commentaries, analyses, and other summary works that place research developments into context should be made immediately available (open access with no embargo period) and reusable via an expansive license such as CC BY.	With millions of research articles published annually, the need for filtering, selection, and curation has never been greater. Commentaries and analyses, including (but not limited to) review articles and research summaries, provide context for the findings described in primary articles. These materials extend the utility of primary research and widen the prospective audience to include policy makers and the general public.	• Commentaries and summaries are an important way for learned societies to add value and continue to earn some subscription income.
Data	Subject to personal privacy, regulatory, and legal restrictions, data underlying specific claims in a research project should be deposited with the necessary metadata into a repository, with efforts taken to maximize findability, accessibility, interoperability, and reuse. Deposits should be made no later than the publication of the first paper based on the data. Data should be considered legitimate, citable products of research.	The independent confirmation of results and conclusions is critical for understanding scientific soundness and informing future research activities. Openly shared data can shed light on negative results and attempted research directions, with the potential to improve efficiency of the research process as well as lead to novel analyses and conclusions.	• Stewardship/ownership of repositories—ensuring these are open and sustainable. • Timing of data release. • Restrictions on data reuse (e.g., text and data mining).

continued

Table Continued

Research Output Type	Exemplar Open Practices	Importance to Open Ecosystem	Concerns/Considerations
Digital Scholarship	Multimedia, digital media, and audiovisual outputs should be made immediately available (open access with no embargo period) and reusable via an expansive license such as CC BY.	Digital scholarship encompasses a range of research outputs in several disciplines (particularly in the humanities). These materials are critical to the scholarly record, particularly when they are made available under a license that permits reuse and remixing.	• Stewardship/ownership of repositories—ensuring these are open and sustainable. • Ensuring that materials are "future proofed" and viable for access and reuse for an extended period of time.
Monographs, Books, Book Chapters, and/or Edited Volumes	All monographs, books, book chapters, and/or edited volumes should be made immediately available (open access with no embargo period) and reusable via an expansive license such as CC BY.	Unrestricted access to, and reuse of, monographs, books, book chapters, and/or edited volumes benefits the research community because it facilitates the discovery of new information, and thus maximizes opportunities for that work to lead to new insights and discoveries.	• Open access for books and longer form content is less developed than journals. Few options/models.
Non-Peer-Reviewed Reports, Posters, and Presentations	All non-peer-reviewed outputs that are appropriate to be shared with the research community (e.g., reports and presentations) should be made immediately available (open access with no embargo period) and reusable via an expansive license such as CC BY.	Unrestricted access to, and reuse of, non-peer-reviewed outputs benefits the research community because it facilitates the discovery of new information, and thus maximizes opportunities for that work to lead to new insights and discoveries.	• Grantees/faculty members may require additional guidance as to what constitutes an appropriate research output.

99

Peer Reviews	Peer reviews should be published with the article (so-called open reports). They can be anonymous or not. The author's response to the reviews should be published as well.	Publishing referee reports makes the process more transparent. Peer reviews contain arguments and ideas that can reveal how thinking in a field evolves. This material should be preserved and made available to others. Additionally, readers have a right to understand the level of scrutiny that a paper has undergone, and it provides them with a window into the editorial process. Because peer reviews are an essential component of the research endeavor, publishing referee reports helps create a pathway for formally crediting this activity.	• Infrastructure limitations. Right now, less than 3 percent of scientific journals allow peer reviews to be published. • Ownership considerations. Who has the right to disseminate referee reports? Authors? Reviewers? Publishers? • Providing credit for peer reviews without compromising anonymity (see ORCID *PLOS*[a] collaboration). • Several initiatives are emerging to support peer-review experiments. For example, ASAPbio has launched ReimagineReview a directory of peer-review trials, inside and outside the journal system.[b]

continued

Table Continued

Research Output Type	Exemplar Open Practices	Importance to Open Ecosystem	Concerns/Considerations
Preprints	Scientists should share preprints (paper drafts that have not yet been peer reviewed for formal publication) by posting in a repository or preprint server that codifies free, unrestricted, and perpetual access to the preprint. Preprints should be posted in a timely manner, ideally at the time of first submission to a journal.	Preprints allow research findings to be quickly and easily available to all and allows researchers to claim priority of discovery, receive community input, and demonstrate evidence of progress for funders and others.	• The growing visibility of preprints may render double-blind peer review more challenging, as prospective referees are exposed to preprints (and their authors) prior to the journal submission and review stages. • Several initiatives are emerging to support preprints. For example, ASAPbio.org is a comprehensive resource for information on preprints, peer reviews, transparency, and so forth. Transpose is a directory of journal policies, co-reviewing, and preprints.[d]

Preregistration Analysis Plans	Indicate in grant proposals, progress reports, and published articles of funded research that the research will be preregistered with an analysis plan. Provide a URL link to preregistration in reports and articles when completed. When results are reported, make a clear distinction between the planned research and any unplanned research or analysis that was conducted. Disclose any deviations from the planned procedures.	Unreported flexibility in data analysis decreases scientific credibility and invalidates common tools of statistical inference. By submitting a detailed study protocol and statistical analysis plan to a registry prior to conducting the work (i.e., preregistering with an analysis plan) the scientist makes a clearer distinction between planned hypothesis tests (i.e., confirmatory tests) and unplanned discovery research (i.e., screening or exploratory research). Preregistration is particularly important for studies that make an inferential claim from a sampled group or population, as well as studies that are reporting hypotheses.
		• May not be appropriate for all types of research, such as studies that do not claim to make inferences, that are purely discovery, that do not test hypotheses, or that generate computational models. • Also see the Registered Reports section, below, as a way to practice prospective registration.
Protocols	Descriptions of the design and implementation of experiments should be made freely available in an open repository that facilitates the sharing, editing, forking (copying and adopting/modifying), and further development. These include study protocols (description of the study plan), and laboratory protocols (detailed description of experimental methods).	Understanding the starting point for work—including assumptions—along with the final study and analysis can provide guidance to other researchers as to additional research avenues to explore. Protocols provide the context to interpret and understand how research results are derived. They can convey exactly what was done and the decisions/compromises that were made on route to a scientific discovery.
		• Protocols can be shared prior to conducting work, which provides insights into research that does not ultimately get published; this is uncommon at present.

continued

Table Continued

Research Output Type	Exemplar Open Practices	Importance to Open Ecosystem	Concerns/Considerations
Registered Reports	Indicate in grant proposals, progress reports, and published articles which parts of the funded research will be submitted as a registered report. In project time-line documentation, add the appropriate time (e.g., 2 to 4 months) for the peer-review process at the beginning of the relevant project phases. Communicate with the funder on any procedural changes that occur as a result of peer-review feedback. If the funder partners with journals to combine reviewer feedback to jointly offer funding and publishing, submit to such solicitations.	Peer review of study protocols with analysis plans, along with dissemination of findings regardless of outcome, addresses publication bias against null results. It also provides the benefits of preregistration by making a clearer distinction between hypothesis tests and discovery research. By submitting funded studies to journals as a registered report, the scientist improves study planning, increases study rigor, and improves scientific credibility. Funders who support this process anticipate that peer-review feedback could change study processes that result in budget changes and are prepared to consider such amendments in response to journal reviewer feedback. Funders can also partner with journals to coordinate review for funding and publishing decisions.	• In some disciplines and in some types of research, infrastructure (including, but not limited to, participating journals) to support registered reports activities is limited. • May not be appropriate for all types of research, such as studies that do not claim to make inferences, that are purely discovery, that do not test hypotheses, or that generate computational models.

Research Materials	Biological and other physical samples (in particular starting materials), research tools (including reagents, animal models, and the like), and other materials (including metadata and identifiers) necessary to reproduce or extend research findings should be made freely available in an open repository no later than the publication of the first paper based on the materials.	Similar to code and data, it allows the independent confirmation of results. Also similar to code and data, broader access to research materials can accelerate research more broadly and allow comparisons across research project or products. Biological materials, such as cell lines, are fundamentally different from data and even software as they may embody a type of "machine" that, through cell expression and the like, can be used to make desirable products, such as a particular valuable protein.	• Cost of maintaining/sharing certain types of samples and quality control. • Stewardship/ownership of repositories—ensuring these are open and sustainable, including detailed descriptions of samples.

continued

Table Continued

Research Output Type	Exemplar Open Practices	Importance to Open Ecosystem	Concerns/Considerations
Theses and Dissertations	All theses and dissertations should be made available (open access with as short an embargo period as possible) and reusable via an expansive license such as CC BY.	Theses and dissertations represent significant contributions to the advancement of knowledge and the scholarly record. The open sharing of these materials offers a particularly unique insight into the research perspective of the emerging generation of scholars.	• Because students often try to publish portions of their theses and dissertations as articles, and because some journals still consider posted electronic theses/dissertations to be "prior publication," a reasonable embargo period may be both necessary and appropriate. • An embargo of substantial length may create an added burden if the author has graduated and left the institution. • Authors of theses and dissertations that disclose a novel process or invention for which a patent may be sought may require longer embargoes. • Some theses and dissertations incorporate works of other copyright owners; this may require additional intellectual property guidance.

[a] See https://theplosblog.plos.org/2019/06/youve-completed-your-review-now-get-credit-with-orcid.
[b] See https://reimaginereview.asapbio.org.
[c] See https://asapbio.org.
[d] See https://transpose-publishing.github.io/#.
NOTE: CC BY – Creative Commons Attribution License.